	Superfamily	Family	Subfamily	Genus	Common name
Simiae	Ceboidea (New World monkeys, platyrrhine monkeys)	Cebidae	Cebinae	*Pithecia*	Saki
				Chiropotes	Saki
				Cacajao	Uakari
				Alouatta	Howler monkey
				Saimiri	Squirrel monkey
				Cebus	Capuchin
				Ateles	Spider monkey
				Lagothrix	Woolly monkey
	Cercopithecoidea (Old World monkeys, catarrhine monkeys)	Cercopithecidae	Cercopithecinae	*Macaca*	Macaque
				Cynopithecus	Black ape
				Papio	Baboon, drill, mandrill
				Theropithecus	Gelada
				Cercocebus	Mangabey
				Cercopithecus	Guenon
				Erythrocebus	Patas monkey (hussar monkey, red monkey)
			Colobinae	*Presbytis*	Langur, leaf-monkey
				Pygathrix	Douc
				Rhinopithecus	Snub-nosed monkey
				Simias	Pig-tailed langur (Mentawi Islands langur)
				Nasalis	Proboscis monkey
				Colobus	Guereza
	Hominoidea (apes and man)	Hylobatidae (lesser apes)		*Hylobates*	Gibbon
				Symphalangus	Siamang
		Pongidae (great apes)	Ponginae	*Pongo*	Orangutan
				Pan	Chimpanzee
				Gorilla	Gorilla
		Hominidae		*Homo*	Man

Note: Names in parentheses are synonyms for the names they immediately follow. Names separated by commas, but not in parentheses, are not synonyms.

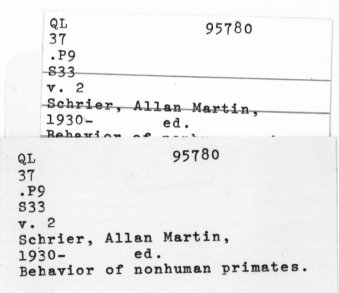

Behavior of Nonhuman Primates
MODERN RESEARCH TRENDS

Volume II

Contributors To This Volume

ROBERT A. BUTLER

ROGER T. DAVIS

ROBERT L. FANTZ

HARRY F. HARLOW

MARGARET K. HARLOW

PHYLLIS JAY

WILLIAM A. MASON

A. J. RIOPELLE

C. M. ROGERS

CHARLES C. TORREY

ROBERT R. ZIMMERMANN

Behavior of
Nonhuman Primates

MODERN RESEARCH TRENDS

EDITED BY

ALLAN M. SCHRIER

PRIMATE BEHAVIOR LABORATORY
WALTER S. HUNTER LABORATORY OF PSYCHOLOGY
BROWN UNIVERSITY
PROVIDENCE, RHODE ISLAND

HARRY F. HARLOW

PRIMATE LABORATORY
DEPARTMENT OF PSYCHOLOGY
UNIVERSITY OF WISCONSIN
MADISON, WISCONSIN

FRED STOLLNITZ

PRIMATE BEHAVIOR LABORATORY
WALTER S. HUNTER LABORATORY OF PSYCHOLOGY
BROWN UNIVERSITY
PROVIDENCE, RHODE ISLAND

Volume II

1965

ACADEMIC PRESS   New York and London

ACADEMIC PRESS, INC.
111 Fifth Avenue, New York, New York 10003

United Kingdom Edition published by
ACADEMIC PRESS, INC. (LONDON) LTD.
24/28 Oval Road, London NW1 7DD

LIBRARY OF CONGRESS CATALOG CARD NUMBER: 65-18435

Second Printing, 1972

PRINTED IN THE UNITED STATES OF AMERICA

To our wives, Judith, Peggy, and Janet

List of Contributors

Numbers in parentheses indicate the pages on which the authors' contributions begin.

ROBERT A. BULTER, Departments of Surgery and Psychology, University of Chicago, Chicago, Illinois (463)

ROGER T. DAVIS, Department of Psychology, University of South Dakota, Vermillion, South Dakota (495)

ROBERT L. FANTZ, Department of Psychology, Western Reserve University, Cleveland, Ohio (365)

HARRY F. HARLOW, Primate Laboratory, Department of Psychology, University of Wisconsin, Madison, Wisconsin (287)

MARGARET K. HARLOW, Primate Laboratory, Department of Psychology, University of Wisconsin, Madison, Wisconsin (287)

PHYLLIS JAY, Department of Anthropology, University of California, Davis, California (525)

WILLIAM A. MASON, Delta Regional Primate Research Center, Covington, Louisiana (335)

A. J. RIOPELLE, Delta Regional Primate Research Center, Covington, Louisiana (449)

C. M. ROGERS, Yerkes Laboratories of Primate Biology, Orange Park, Florida (449)

CHARLES C. TORREY,[*] Department of Psychology, Cornell University, Ithaca, New York (405)

ROBERT R. ZIMMERMANN, Department of Psychology, Cornell University, Ithaca, New York (405)

[*] Present Address: Lawrence College, Appleton, Wisconsin

Preface

Research on the behavior of nonhuman primates has mushroomed in the last decade. Not many years earlier, primate behavior was a major area of study in only two laboratories in the United States, and only a few isolated field studies had been made. Today, in addition to the pioneering Yerkes Laboratories of Primate Biology and the University of Wisconsin Primate Laboratory, there are primate behavior laboratories in many university psychology departments and government research institutions, and long-term field studies have become almost common. But the wealth of new data on primate behavior is relatively inaccessible to most behavioral scientists. It cannot be described adequately in books covering the wider field of animal behavior or comparative psychology, and the original research reports are scattered through hundreds of volumes of scientific journals.

To make a substantial part of this new knowledge of primate behavior more readily available, we have tried to obtain thorough and up-to-date descriptions of fifteen important research areas. The resulting volumes do not include every study in the literature on primate behavior; rather, each of the chapters attempts to give a coherent, integrated view of its particular area. Some historical background is provided, but the emphasis is on modern research trends. We hope that these chapters will provide new perspectives not only to researchers using nonhuman primates as subjects, but also to other readers—for example, to those who are mainly interested in human behavior but are also curious about man's closest relatives, and to those who study rat behavior but wonder how well rodent results (and rodent theories) can be extrapolated to primates. Studies of nonhuman primates have already been influential in the conceptual development of the behavioral sciences; this influence may well increase as work on primate behavior becomes better known.

We have assumed that many readers would be familiar with the study of behavior, but not necessarily with concepts and techniques specific to primate research. Each chapter either describes techniques and problems fully enough for nonspecialists or provides cross-references to detailed descriptions in other chapters. In addition, Volume I, which concentrates on studies of learning and problem-solving, starts with a chapter that emphasizes methods used in many of these studies.

Nonhuman primates are referred to by their vernacular names in this book. Identification of some animals presents special problems, which are discussed further in the Appendix, but we may note here that the scientific name is also given in each chapter the first time that a particular vernacular name is mentioned. A classification of living primates is presented on the endpapers of each volume. Although very little is said about the prosimians in this book, we believe that this gap in our knowledge is temporary and that the approaches to behavior that have been so fruitful with apes and monkeys will soon be extended to the prosimians.

In acknowledging our gratitude to others, we recognize immediately our double debt to the authors, who literally made these volumes possible. We thank them first for their contributions, which reflect their high competence as researchers and scholars, and second for their cooperation with editors who thought that their specialties might be of interest to nonspecialists and tried to edit accordingly. We are also grateful to a number of other people: Harold Schlosberg, the late chairman of the Brown University Psychology Department, who was an encouraging and understanding friend; other members of the Department, who provided valuable advice on several questions related to their special areas of interest; Kathryn M. Huntington, who tirelessly typed and retyped manuscripts in various stages of revision; Richard A. Lambe and Karen E. Lambe, who were an excellent proofreading team; Vicky A. Gray, who gave editorial assistance in a variety of ways; and Kenneth Schiltz, who labored through the University of Wisconsin's photographic files to find figures for several chapters. Preparation of this book was supported in part by U. S. Public Health Service Grant MH-07136, from the National Institute of Mental Health.

April, 1965   ALLAN M. SCHRIER
                                                 HARRY F. HARLOW
                                                 FRED STOLLNITZ

Contents

Chapter 8

THE AFFECTIONAL SYSTEMS

By Harry F. Harlow and Margaret K. Harlow

Chapter 9

DETERMINANTS OF SOCIAL BEHAVIOR IN YOUNG CHIMPANZEES

By William A. Mason

Chapter 10

ONTOGENY OF PERCEPTION

By Robert L. Fantz

Chapter 11

ONTOGENY OF LEARNING

By Robert R. Zimmermann and Charles C. Torrey

Chapter 12

AGE CHANGES IN CHIMPANZEES

By A. J. Riopelle and C. M. Rogers

Contents of Volume I

Chapter 8

The Affectional Systems[1]

Harry F. Harlow

Department of Psychology Primate Laboratory and Wisconsin Regional Primate Research Center, University of Wisconsin, Madison, Wisconsin

and

Margaret K. Harlow

Department of Psychology Primate Laboratory, University of Wisconsin, Madison, Wisconsin

I. INTRODUCTION

One of the many ways in which primates differ from other animals is in the strength, duration, and diversity of their affectional systems, systems which bind together various individuals within a species in coordinated and constructive social relations. Five affectional systems may be described for most primate forms. Each system develops through its own maturational stages and differs in the underlying variables which produce and control its particular response patterns. Typically the maturational stages overlap rather than being discrete during the course of

[1] This research was supported by grants MH-04528 and FR-0167 from the National Institutes of Health, U. S. Public Health Service, to the University of Wisconsin Primate Laboratory and the Wisconsin Regional Primate Research Center, respectively, and by funds from the University of Wisconsin Graduate School.

their development. These five affectional systems, in order of development, are: (1) The infant-mother affectional system, which binds the infant to the mother; (2) the mother-infant or maternal affectional system; (3) the infant-infant, age-mate, or peer affectional system through which infants and children interrelate with each other and develop persisting affection for each other; (4) the sexual and heterosexual affectional system, culminating in adolescent sexuality and finally in those adult behaviors leading to procreation; and (5) the paternal affectional system, broadly defined in terms of positive responsiveness of adult males toward infants, juveniles, and other members of their particular social groups. Although monkeys do not live in family groups, field studies make it quite obvious that a paternal affectional system exists throughout all or at least most members of the primate order (see Chapter 15 by Jay).

The number of affectional systems in the primates can be expanded to an almost unlimited degree according to the whim of any man who classifies, and we could, for example, have an adult female-female affectional system and an adult male-male affectional system. However, for purposes of simplicity of classification, we believe that our fivefold postulated affectional systems will serve as an effective basis for behavioral and psychophysiological investigation. Systematic papers concerning the infant-mother system (Hansen, 1962; Harlow, 1959; Rosenblum, 1961), the mother-infant affectional system (Hansen, 1962; Harlow *et al.*, 1963), the infant-infant affectional system (Harlow, 1962a), and the heterosexual affectional system (Harlow, 1962c; Harlow & Harlow, 1962) in rhesus monkeys (*Macaca mulatta*) have already been presented.

II. THE INFANT-MOTHER AFFECTIONAL SYSTEM

The infant-mother affectional system is enormously powerful and probably less variable than any other of the affectional systems. It is not surprising that this is so, because strong infant-mother ties are essential to survival, particularly in a feral environment. This system is so binding that many infants can survive relatively ineffective mothering, and the system will even continue with great strength in the face of strong and protracted punishment by unfeeling mothers.

A. Development

The infant-mother affectional system develops through four stages: (1) a reflex stage, (2) a comfort and attachment stage, (3) a security stage, and (4) a separation stage. The nature of the infant's affectional responses to the mother throughout all developmental stages is a function of many variables, including the species (see DeVore, 1963; Jay, 1963), the nature of the mother, the nature of the total environment, and individual, inherent infant differences. Because of this wealth of variables, it is impossible to assign exact days or weeks when one developmental stage ends and another begins. Stages gradually blend from one to another; transitions are not made in hours or days but in weeks or months; stages are commonly characterized by overlapping rather than discrete behaviors, and all stages, with the possible exception of the reflex stage, can be extended or abbreviated by special experimental procedures.

1. REFLEX STAGE

In rhesus monkeys, the reflex stage persists during the first 15 to 20 days of the infant's life, depending on the reflex selected and the maturity of the infant at birth. There are two families of reflexes, one associated with nursing and the other with intimate physical contact, and both of these response systems are essential to the early survival of the infant. The most fundamental reflex associated with nursing is the rooting reflex. Stimulation of the baby's face, particularly near the mouth, causes the infant to rotate its head either sideways or vertically until the nipple is touched. The nipple is then orally engulfed and sucking responses follow (Fig. 1). Mother monkeys and humans can also elicit this reflex by a very specific auditory stimulus of low intensity produced by rapid tongue-teeth engagements and disengagements. This stimulus causes the infant to rotate the head toward the source even though neonatal monkeys are relatively unresponsive to most auditory stimulation. Another reflex that may be related to the act of nursing is that of forced upward climbing. If a neonatal monkey is placed on a wire ramp, it will climb up the ramp and even climb over the end of the ramp and fall to the floor unless it is restrained. This fits our criteria of involuntary reflexes, which are elicited behaviors continuing even when entirely inappropriate to the total environmental situation. We hypothesize that upward climbing may assist the infant in attaining the breast by climbing up the mother's body until the rooting reflex enters or the baby is restrained by the mother's arms at the level of the breast.

There is also a family of reflexes which guarantee that the infant will maintain intimate contact with the mother's body. There is a basic clinging reflex which leads the neonate to attach by its arms and legs intimately to the ventral surface of the mother and orally to the mother's breast. This is a reflex of enormous strength and power in the infant monkey and is doubtless essential to its survival. The neonatal infant must respond cooperatively to the mother monkey, which could not give the baby continuous total physical support and still keep up with the daily peregrinations of the monkey group, often entailing several miles of travel a day (see Chapter 15 by Jay). Dual separation from the group would be catastrophic to both mother and infant.

The power of the clinging reflex in neonatal rhesus monkeys can be illustrated by placing an infant on its back on any flat, solid surface. The infant immediately rotates into a prone position through the operation

Fig. 1. Sucking response by 3-day-old rhesus monkey.

of the basic and powerful righting reflexes. However, if the infant rhesus is similarly placed while clinging to a cylinder, the infant continues to lie passively on its back (Mowbray & Cadell, 1962). In the neonate, the power of clinging is so great as to be entirely prepotent over the righting reflex.

The grasp reflex is no doubt another member of this reflex family. If the palmar surface of the hand or the plantar surface of the foot is stimulated by a thin rod, the forced grasp reflex is elicited so strongly that the animal may be suspended in air (Mowbray & Cadell, 1962). This reflex is without doubt an important mechanism ensuring infant-mother attachment at the reflex stage. As is well known, the grasp reflex can be elicited also in many neonatal humans.

The reflex stage comes closer than any other stage to having a sharp temporal cutoff. But even at this developmental level, maturational differences exist in infant monkeys. Many basic reflexes appear and subsequently fade at different times and in different sequential orders, and are replaced by more subtle responses giving the infant greater freedom for reactivity.

By describing a reflex stage of infant-mother affectional development we do not mean to imply that all infant attachments during the first 20 days of life are determined by forced reflex responses. During the tenth to twentieth day of life, the rhesus monkey develops effective locomotor capabilities and may actually break physical contact with the mother for brief periods of time, both in the wild and in the laboratory in our playpen situation (see Section II, A, 3). Day by day the reflex pattern of dependence gradually subsides and is supplanted by voluntary activities.

Furthermore, learned infant-mother attachments are established during this interval. The infant monkey can acquire conditioned responses during the first 5 days of life and learned discriminations from 10 days onward (see Chapter 11 by Zimmermann and Torrey), and an animal that can learn will learn. There is every reason to believe that the infant learns attachments to a specific mother (*the* mother) long before our postulated reflex stage has passed, has acquired rudimentary visually-controlled imitative responses, and has learned to respond to maternal gestures inviting contact or warning of danger. Unfortunately, definitive experimental data do not exist, but ample observational data from laboratory- and feral-raised monkeys give full support to this position.

2. STAGE OF COMFORT AND ATTACHMENT

As the reflex stage gradually wanes, it is replaced by a stage of comfort and attachment in which the infant receives little from the mother, other than the satisfaction of its basic bodily needs, nursing, contact, warmth, and probably proprioceptive stimulation and protection from danger. This stage in the monkey may be relatively brief, ending at about 60 to 80 days, but in the human it probably persists for 6 to 8 months or more. During this stage the mother is highly protective (Section III, A, 1), but the infant receives only a moderate sense of security from the mother and expresses little lack, or only transient lack, of security in her absence.

During the stage of comfort and attachment, the infant rhesus monkey stays in close proximity to its mother. Tight reflex clinging gradually diminishes, and the infant monkey spends a considerable amount of time either loosely cradled in the mother's arms, exploring her body, or exploring the adjacent physical world (see Section III, A, 1).

The intimate physical responsiveness of the infant macaque to the mother is illustrated by the time spent on the body of an inanimate cloth-covered surrogate mother. Eight infant monkeys were raised with both a cloth surrogate and a wire surrogate always available. Four of the infants were fed on demand by the cloth surrogate and four by the wire surrogate, but this difference turned out to be relatively unimportant. As shown in Fig. 2, each group of four infants averaged

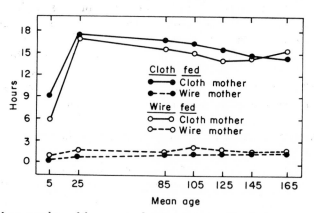

FIG. 2. Mean number of hours per day in contact with cloth versus wire mother surrogates. (After Harlow & Zimmermann, 1959.)

over 15 hours a day in contact with the cloth surrogate. Much of this time was devoted to ventral clinging, a response which normally disappears with the stage of comfort and attachment, but which persists in this situation long after the existence of the stage of security can be experimentally demonstrated. The explanation of this peculiar situation lies in the nature of the artificial cloth surrogate, which, from an ethological point of view, was doubtless not only a good but a "supernormal" mother in that it was always available, never moved away, and never punished.

Although the attainment of primary satisfaction from nursing and physical contact characterizes the stage of comfort and attachment, we do not believe that these contribute equally to the resolution of this stage and the transition to the next stage. We strongly believe that intimate physical contact is the variable of primary importance in enabling the infant to pass from the stage of comfort and attachment to a stage of security, specific security from a specific maternal figure (Harlow, 1958). Thus, the stage of comfort and attachment is a stage during which the infant establishes a bond of maternal trust, which it later uses in forming and regulating broader behavioral patterns.

3. Stage of Security

The development of a security stage was first brought to our attention during a test of fear in the home cage. A mechanical bear or a mechanical dog was presented suddenly by raising an opaque screen and exhibiting one of the monsters in an adjacent chamber. After 20 days of age, the infants raised with both cloth and wire surrogates usually fled from the monster and clung intimately to the cloth mother, a natural response to fear. From 80 days of age onward, this act of intimate attachment led the infants to relax quickly, and then, within the brief 15-second test period, they often actually left the mother, approached and even manually explored the bear or dog. We were not surprised to find that infant monkeys would attach to a cloth surrogate and be comforted by her but we were surprised that an inanimate dummy mother could impart feelings of security to her infant at an appropriate developmental age.

We performed a similar series of open-field experiments, comparing monkeys raised on mother surrogates with control monkeys raised in a wire cage containing a cheesecloth blanket from days 1 to 14 and no blanket after that. The infants were introduced into the strange environment of the open field, which was a cubical room measuring 6 feet on each side, containing many stimuli known to elicit curiosity-manipulatory responses in baby monkeys. Infants raised with single mother surrogates were placed in this situation twice a week for 8 weeks, no mother surrogate being present during one of the weekly sessions and either a cloth or a wire surrogate (the kind that the experimental infant had always known) being present during the other sessions. Four infants raised with dual mother surrogates and four control infants were subjected to similar experimental sequences, the cloth mother being present on half of the occasions.

As soon as they were placed in the test room, the infants raised with cloth mothers rushed to their mother surrogate when she was present and clutched her tenaciously, a response so strong that it can only be adequately depicted by motion pictures. Then, as had been observed in the fear tests in the home cage, they rapidly relaxed, showed no sign of apprehension, and began to demonstrate unequivocal positive responses of manipulating and climbing on the mother. After several sessions, the infants began to use the mother surrogate as a base of operations, leaving her to explore and handle a stimulus object and then returning to her before going to a new plaything. Some of the infants even brought the stimulus objects toward the mother. The behavior of these infants changed radically in the absence of the mother. Emotional indices such as vocalization, crouching, rocking, and sucking increased sharply. Typical response patterns were either freezing in a crouched position (Fig. 3) or running around the room on the hind feet, clutching themselves with their arms. Although no quantitative evidence is available,

Fig. 3. Four-month-old rhesus monkey freezing in open field in absence of mother surrogate. (From Harlow, 1958.)

contact and manipulation of objects was frenetic and brief, as opposed to the playful type of manipulation observed when the mother was present.

In the presence of the mother, the behavior of the infants raised with single wire mothers was both quantitatively and qualitatively different from that of the infants raised with cloth mothers. Not only did most of these infants spend little or no time touching their mother surrogates, but the presence of the mother did not reduce their emotionality. These differences are evident in the mean number of 15-second periods spent in contact with the respective mothers (Fig. 4) and in a composite emotional index for the two stimulus conditions (Fig. 5). Although the infants raised with dual mothers spent considerably more time in contact with the cloth mother than did the infants raised with single cloth mothers, their emotional reactions to the presence and absence of the mother were highly similar, the composite emotional index being reduced by almost half when the mother was in the test situation. The infants raised with wire mothers were highly emotional under both

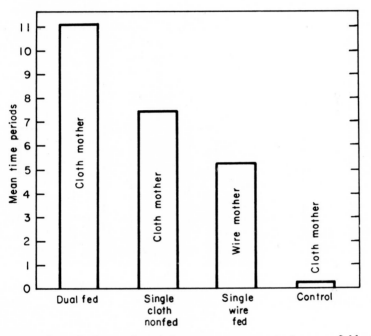

Fig. 4. Number of 15-second periods spent on surrogate in open field. (After Harlow & Zimmermann, 1959.)

conditions and actually showed a slight, though nonsignificant, increase in emotionality when the mother was present. Although some of the infants raised with a wire mother did touch her, their behavior was similar to that observed in the home-cage fear tests. They did not clutch and cling to their mother as did the infants with cloth mothers; instead, they sat on her lap and clutched themselves, or held their heads and bodies in their arms and engaged in convulsive jerking and rocking movements similar to the autistic behavior of deprived and institutionalized human children. The lack of exploratory and manipulatory behavior of the infants raised with wire mothers, both in the presence and absence of the wire mother, was similar to that of the infants raised with cloth

mothers when the mother was absent. Such contact with objects as
was made was, brief, erratic, and frantic. None of the infants raised
with single wire mothers played persistently and aggressively, as did
many of the infants that were raised with cloth mothers.

Four control infants, raised in bare-wire cages without a mother surrogate, had approximately the same emotionality scores when the mother
was absent that the other infants had in the same condition, but the
control subjects' emotionality scores were reliably higher in the presence

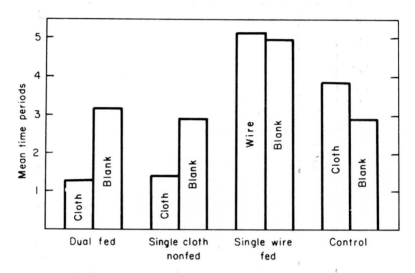

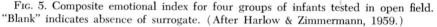

Fig. 5. Composite emotional index for four groups of infants tested in open field.
"Blank" indicates absence of surrogate. (After Harlow & Zimmermann, 1959.)

of the mother surrogate than in her absence. This result is not surprising,
since recent evidence indicates that the cloth mother with the highly
ornamental face becomes an effective fear stimulus, for monkeys that
have not been raised with her, when fear of specific external objects becomes enhanced between 60 and 100 days of age.

There can be no doubt that real monkey mothers play as great a role
as cloth surrogate mothers do (and probably a far greater role than surrogates) in imparting security to their infants. Effects on behavioral development of being raised by real monkey mothers and by cloth surrogate mothers were directly compared in our playpen test situation,
which consists of large living cages each housing a mother and an infant
and adjoining a compartment of the playpen (Fig. 6). A small opening
in each living cage restrains the mother, but gives the infant continuous
access to the adjoining playpen compartment.

In two playpen situations babies were housed with their real mothers, and in a third setup the babies were housed with cloth mothers. During two daily test sessions, each an hour long, the screens between playpen compartments were raised, letting the infant monkeys interact as pairs during the first 6 months and as both pairs and groups of four during the ensuing 9 months. Two experimenters independently observed and recorded the behavior exhibited during test sessions.

This test situation was used primarily to observe maternal behavior and age-mate affectional development, and no attempt was made to trace experimentally the development of the stage of security. Casual

PLAYPEN UNITS

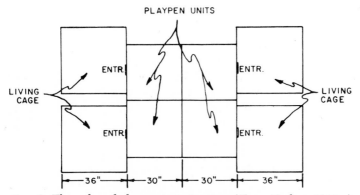

Fig. 6. Floor plan of playpen test situation. (From Harlow, 1962a.)

observation, however, indicated that the infants used both the real mothers and the surrogates as security sources. During the observation periods, the infants would break up bouts of play, rush back to the mother and make physical contact, and then return to the playpen units to explore and play with their partners, with physical objects, or with both. The security bond between the infants and their real mother was more precise and possibly more intense than was the security bond between infants and the cloth surrogates. When the infants with real monkey mothers were frightened, they always returned to their own mothers, while the infants with cloth surrogates often went to an inappropriate mother when frightened. Two or even three infants would sometimes run and attach to a single cloth surrogate.

Field studies clearly demonstrate that infant monkeys in the wild use their mothers as security figures. During the first year of life, perhaps two years in the case of baboons (*Papio*), the infants leave play bouts and exploratory periods to return to their mothers. After being reassured by physical contact, or even token physical contact, they leave the mothers' physical domain to reestablish their diverse exploratory or play behaviors with age-mates (DeVore, 1963).

The development of the security stage of infants in the wild is a function not only of the presence of the mother but also a function of the social role of the mother. If the infant is fortunate enough to have a highly dominant female as his mother, he has little or nothing to fear since he can treat other infants and even other mothers with impunity. If an infant is unfortunate enough to be the offspring of a low-dominant mother, he must walk with caution and watch with care since his mother can in no sense guarantee security from other infants or from other mothers (Imanishi, 1957).

4. SEPARATION STAGE

The dissolution of the affectional relationships between a particular infant and its mother is in part a function of the infant's behaviors, directly or indirectly. The outer-world lures of exploration and play are powerful forces acting to produce part-time maternal separation. However, these forces do not appear to be adequate, in and of themselves, to break the infant-mother bond. Hansen (1962), studying the relations between infants and real mothers in the playpen situation, pointed out that there was no true infant-mother separation within the period of his study. Behaviors that form the basis of this stage were observed in the development of maternal ambivalence (Section III, A, 2), but the mother-infant bond, especially the psychological bond, persisted through 21 months, at which time the study ended. Thus, when frightened, the infants always returned to their mothers and were generally accepted by them. At night, the infants were always seen in contact with their mothers, cradled in their arms or lying on their bodies. Indeed, it is entirely possible that true rejection and mother-infant separation would never have occurred within the social structure of the playpen group. In the case of the male infants, mother-infant affectional responding might well have come to be gradually replaced by elements of the heterosexual affectional system.

Even greater resistance to a separation stage was exhibited by monkeys raised in the dual-surrogate situation during the first 180 days of life and then separated. Retention of the affectional responses was tested during the first 9 days of separation, then at six successive 30-day intervals, and then at four intervals of 90 days each, by measuring how long the monkeys spent in contact with each mother surrogate (Fig. 7). Obviously, the affection of the infants for the cloth mother surrogate was extremely persisting.

All of these data attest to the enormous power and persistence of the infant-mother bond. It is easy to see how this bond, in both rhesus monkeys and human beings, can become so intimate and prolonged as to seriously disrupt other affectional relations, such as those between peers. However, the fact remains that eventual physical and psychological separation is the general rule in all primate species, and we will return to the problem when we discuss the maternal or mother-infant affectional system.

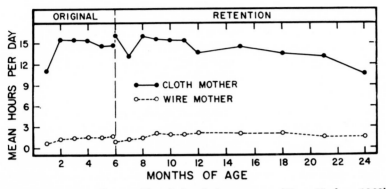

Fig. 7. Retention of affectional bonds for cloth surrogate. (From Harlow, 1962b.)

B. Variables Determining Infant-Mother Affection

The variables operating during the infant developmental stages of comfort, attachment, and security have already been described in detail (Harlow, 1959; Harlow & Zimmermann, 1959). Our data suggest that clinging is the primary variable underlying infant-mother affection but that many other variables operate as well. These include nursing, warmth, and proprioceptive stimulation. Without doubt, there are also visual and auditory variables, both unlearned and learned, but we have not yet subjected these to detailed analyses.   .

III. THE MATERNAL AFFECTIONAL SYSTEM

A. Development

The mother-infant or maternal affectional system also develops through a series of orderly stages involving both maturation and learning. As we have pointed out in the case of the infant-mother affectional system, transition from one stage to another never occurs during any brief critical period measured in terms of hours or days but is a gradual and continuous process and the stages are characterized by overlapping development rather than by discrete development.

Within these limitations, the mother-infant affectional system is best described in terms of three stages: (1) maternal attachment and protection, (2) maternal ambivalence or transition, and (3) maternal separation and rejection. Most of our data on the development of the maternal affectional system were obtained in our playpen test situation (Section II, A, 3). Details of the experimental procedures and definitions of the items that were scored have been described elsewhere (Hansen, 1962; Harlow *et al.*, 1963).

1. STAGE OF MATERNAL ATTACHMENT AND PROTECTION

The stage of maternal attachment and protection, in the case of most mother monkeys, begins at the time of birth or within a few minutes after the baby is born. This stage is characterized by maternal responses which are almost totally protective, including cradling, nursing, grooming, bodily exploration, restraining the infant when it attempts to leave, and retrieving the infant when it does escape. The developmental course of intimate infant-mother bodily contacts as described by Hansen (1962) is illustrated in Fig. 8. Ventral clinging was at first very frequent, but became progressively less frequent from the first to the third month. Cradling responses followed a very similar course. Clinging and cradling are essentially identical patterns, one initiated by the infant, the other by the mother. Gradually the cradling pattern changes to a maternal pattern of loose physical attachment in which the infant monkey is not closely attached to the mother's ventral surface but sits on or near her haunches, with the mother's arms touching the infant's body or being easily available. This change from close cradling to loose attachment is primarily a function of the maternal role—if two neonatal rhesus monkeys

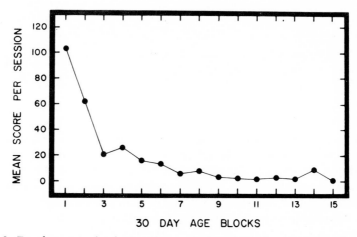

FIG. 8. Development of infant-mother ventral contact. Scores are number of 15-second intervals in which ventral clinging was observed during 1-hour sessions. (After Harlow *et al.*, 1963.)

are placed together, intense ventral clinging occurs (Fig. 9) and increases progressively for the first 90 days or more. Infant-infant ventral clinging is a pattern inimical to effective infant-infant affectional development since it seriously inhibits the infant's play behaviors. Thus the mother serves an important balancing role in providing her baby with enough contact and comfort to impart a sense of security but denying a physical-attachment bond so strong as to impair effective intercourse by the infant with members of its age and peer groups.

FIG. 9. Intense ventral clinging by two 5-month-old macaques placed together at birth.

Nutritional and nonnutritional contacts with the nipple followed a similar developmental course (Fig. 10). All of these data partially reflect the fact that playful interaction was permitted only during the 2 hours of measurement each day. The infants took full advantage of the opportunity to interact, and this undoubtedly obscured the full extent of maternal attachment. Limited observations of mothers and infants when the infants were not permitted to interact with each other support this view. During the early morning and late evening, all mother-infant contacts, and particularly nipple contacts, were enhanced.

Another measure of the development of the stage of maternal attachment and protection was that of restraining and retrieving responses. These responses increased during the first 40 days of life and then decreased rather abruptly. No doubt such measures reflect in part the greater mobility of the infant and increasing interest in exploring the playpen units and interacting with other infants. However, they also represent a decrease in maternal concern. When infants first attempted to leave the living cages the mothers prevented all egress, but the mother monkeys then went through a period in which they appeared to agonize

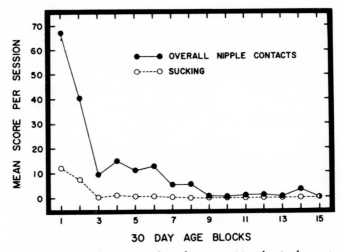

Fig. 10. Development of nutritional and nonnutritional nipple contact. (After Harlow *et al.*, 1963.)

over whether their infant could leave or could not leave. Finally the mothers left the choice almost entirely to the infant, intervening only when disturbed by some fear-evoking stimulus such as a sudden movement by an experimenter or a noise outside the test room. Field studies demonstrate that this period of increasing separation between mother and infant is a function of the total environment, including nature and number of other infants, dominance role of the mother, and species differences (Jay, 1963; Washburn & DeVore, 1961). However, we suspect that early mother-infant separation was enhanced by the playpen situation, since the mother could not follow and retrieve her infant once it had entered a playpen unit. The infants frequently entered these units and remained there undisturbed or seldom frightened, which should lead to experimental extinction of maternal concern.

Maternal protection undoubtedly persists longer and with greater intensity than does maternal attachment. Long after attachment responses have become the primary prerogative of the infant, the mother macaque will rally to protect her infant against any danger, real or imagined. A mother monkey that has become relatively indifferent to her infant changes into an aggressive and bellicose female at the appearance of any danger signal. These responses can be aroused, in both the laboratory and the feral environment, until and perhaps after total physical separation between mother and child has been attained.

Another measure of maternal attachment and protection is that of responses of mother monkeys to infants of other mothers ("other infants"). Despite considerable variability among mothers, the developmental pattern appears to mimic that of the relation between mothers and their own infants, but with great temporal abbreviation. Many mother monkeys show attachment and protection toward other infants during the first few days or even weeks after birth of their own infants if other infants are made directly available. Initial responses to other infants first occurred from 19 to 27 days postpartum in our playpen situation. The initial maternal responses were almost entirely positive but rapidly changed to ambivalent responses and then to aggressive, punitive behaviors, i.e., total and complete separative behaviors.

Strong maternal responsiveness by mature nonpregnant female monkeys has been frequently observed in both laboratory (Rosenblum, 1961) and feral conditions (DeVore, 1963, p. 313; Jay, 1963, p. 289). Furthermore, strong maternal behavior has been observed in prepubertal female monkeys. Preadolescent females, indeed all females, show a deep and pervading interest in newborn infants and a compulsive desire to make gentle physical contact with them whenever this is possible. Thus, a preadolescent female monkey, even though enormously frightened by a mother because of differential dominance status, will frequently sidle up to the mother with the face averted, a common monkey posture of deference, attempt to make contact with the mother's body and gradually edge a hand between the mother's body and the infant's body. Furthermore, preadolescent females readily assume adult-type maternal responses to infants as soon as the mother permits the infant to be taken for brief periods of time. There appear to be species differences, both in the willingness of the mother to allow handling by other females and probably in the intensity of the interest shown by preadolescent and other females (see Chapter 15 by Jay). The greatest freedom of infant handling that has yet been reported was observed in field studies of langurs (*Presbytis entellus*) by Jay (1963). Because of this freedom of infant handling in some monkey species, infant bonnet macaques (*Macaca radiata*) and particularly infant langurs enjoy a considerable amount of mothering by "aunts" as well as by their own mothers. This is a problem that has not been given detailed laboratory investigation, but it is clear that the behaviors of maternal attachment and protection occur in females other than the biological mother.

Motherhood

> If you're endowed with motherhood
> Babies feel good as babies should.
> And you will wish and try to win
> The right to touch a baby's skin.

2. Transitional or Ambivalence Stage

As the initial stage of attachment and protection wanes, there develops a transitional or ambivalent maternal stage characterized by increasing frequency of indifference and, from time to time, comparatively harsh punishment by the mother of her own infant. Punishment of her own infant by the rhesus mother begins to develop toward the end of the third month and reaches a maximum during the fourth and fifth month, with a subsequent decrease.

The development of the various phases of this transitional or ambivalent maternal affectional stage was arbitrarily plotted in Fig. 11 in terms of an over-all index of maternal responsiveness to both own and other infants. Positive responses to other infants predominated during the first month, positive and negative responses were essentially equal during the second month, and negative responses predominated thereafter. Thus, if one arbitrarily takes 50% as a measure of maternal ambivalence and some arbitrary value (e.g., 10%) for separation, ambivalence appeared during the second month toward other infants and not

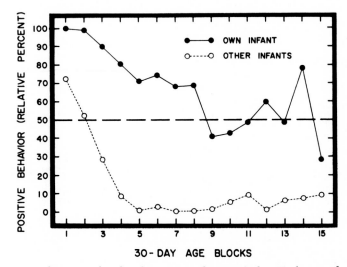

Fig. 11. Development of ambivalent maternal stage. (After Harlow *et al.*, 1963.)

until the ninth month toward the mothers' own infants. Separation appeared at 4 months for other infants and did not appear for the mothers' own infants during the 15 months of development included in Fig. 11.

During the transitional maternal stage, the mother's rejecting responses became progressively harsher, consisting of vigorously shaking the infant from her body or even stiff-arming the infant when it attempted to initiate bodily contact. There were great individual differences among mother monkeys in the development and intensity of the negative, punishing responses.

3. STAGE OF MATERNAL SEPARATION OR REJECTION

Although we observed two groups of four mother-and-infant pairs of rhesus monkeys for 21 and 18 months, we never saw complete separation of any mother-infant pair (see Section II, A, 4). We believe that this was an artifact of our experimental situation. Even though the playpen was a king-size apparatus and occupied most of a 14-feet-square room, it was a situation conducive to the artificial maintenance of mother-infant bonds. The infants were never free to escape from all visual and auditory contacts with their mothers nor to associate with age-mate playgroups independently of some maternal supervision. The field studies reported by DeVore (1963, pp. 327–328) on the olive baboon (*Papio doguera*) and Jay (1963, pp. 296–300) on the langur indicate that the primary variable conducive to total physical separation of mother and child is the advent of the next baby. We made no effort to breed our macaque mothers during the course of this experiment, so there were no new babies. It is obvious that different experimental situations and experimental designs could circumvent these limitations of our present exploratory, normative studies.

B. Variables Influencing Maternal Behavior

The rather high variability of all of our measures of maternal affection indicates that there are many variables influencing maternal behavior of monkeys. Many of these variables appear to be similar or complementary to the variables that regulate the infant-mother affectional system. However, the relative importance of these variables doubtlessly differs between the mother-infant and infant-mother systems, and there are other variables that appear specific to the maternal affectional system.

Obviously the operation of the maternal affectional system involves external stimulation, experiential variables, and many hormonal forces. External stimuli include physical contact, warmth, sucking, and visual and auditory cues provided by both the infant and the physical environment. Experiential variables involve the mother's early life history, interaction with each individual infant, and experiences with each successive infant she bears. Unquestionably, maternal behavior is influenced to some degree by hormonal variables relating to pregnancy and parturition and the resumption of the normal ovulatory cycle.

1. External Stimulus Variables

Intimate clinging by the infant appears to be an important stimulus eliciting the maternal affectional response. As the infant becomes progressively more independent and spends less time intimately physically related to the mother, the mother macaque's affectional feedback wanes and the shift from the stage of attachment and protection to that of ambivalence develops. Unfortunately, simple observation of mother-infant interactions in the playpen situation does not make it possible to establish cause-and-effect relations. Change in maternal interest may be an effect of decreasing infant contact or decreasing infant contact may be an effect of waning maternal interest. Both variables may be operating simultaneously, or both may reflect the operation of other variables as yet undiscovered.

The importance of infant clinging is suggested by the fact that the mother macaque physically separates herself from the previous infant when the stimulation of intimate ventral neonatal contact is reinstated by the birth of a new infant. All physical contactual ties and all acquired experiential variables associated with the older infant are then overridden. Even if the older infant remains in close proximity to the mother and neonate, which is a common observation, he is physically disassociated from the mother, although grooming has been reported between mothers and older siblings after the birth of a new baby (see Chapter 15 by Jay).

Such observations give no answer to the question of the relative importance of the clinging and nursing variables in the elicitation of maternal responsiveness, but some information has been obtained in experiments involving either adoption or temporary separation. We have observed a mother rhesus monkey whose baby was removed at birth and the mother given the opportunity to adopt a kitten. The adoption was complete as long as the mother clung to the kitten and successful nursing was initiated. But the kitten could not cling to the mother and as a result frequently fell to the floor of the cage when the mother moved about. For several days the mother repeatedly retrieved the kitten, then gradually lost interest and in a few days abandoned the adoptee. Nursing may be an important variable, but nursing alone is not sufficient to maintain maternal affection.

We know that clinging by an infant monkey may elicit and maintain maternal affectional responses in nonlactating females. The infants of two multiparous rhesus monkeys were removed at birth; 4 and 9 months later, respectively, infant monkeys 78 and 38 days old were placed with them. In the former case the infant quickly attached to the adult female and normal patterns of motherhood quickly developed and were maintained. In the latter case the adult female initially ignored the infant, which alternately cried and shrieked. After a period of time this adult female, sitting on a perch off the floor, watched the baby intensely and suddenly dropped to the floor, picked up the infant, and did not release it for 3 days. Although early experience variables may have been operating, we believe that infant-mother contacts were basic variables maintaining maternal responsiveness in these nonlactating females. Eventually the mothers did lactate and apparently provided biochemically-normal milk (Harlow *et al.*, 1963).

A single case in which attempted adoption failed was equally informative. The baby was separated from its mother at birth and developed a bizarre, autistic pattern of self-clutching, rocking, and penis-mouthing when caged alone for 38 days. When placed with a mother whose own baby had been removed at 34 days of age, the autistic infant folded into a tight ball and screamed when touched by the mother. The mother vacillated between approach and avoidance, made some body, hand, and grooming contacts during a 4-day period, and then, except for occasional grooming, abandoned the totally unresponsive infant. The data suggested that visual and auditory variables of an infant monkey elicited maternal affectional responses but that these never culminated in adoption because the infant made no effort to contact or cling to the mother. That the infant, not the mother, was to blame became apparent when this same mother adopted a congenitally blind baby that had been mercilessly abused for 2 weeks by its own mother—one of our so-called "motherless mothers" (Section III, B, 2).

Renewed maternal responsiveness following clinging by infants separated from their mothers for 3 weeks at approximately 180 days of age has been reported by Seay *et al.* (1962). Separation was accomplished by inserting Plexiglas partitions between the living cages and the play cells of the playpen apparatus. In three of the four cases the infants immediately re-established ventral-ventral clinging and were immediately accepted by their mothers. The fourth infant was unable to "retrieve" an indifferent mother for over 24 hours but succeeded later in making ventral-ventral contact and was then accepted totally by the adult female.

A second study on rhesus monkey mother-infant separation has given similar data (Seay & Harlow, in press) even though the mothers were removed from the room for a 2-week separation period. The effects of this total separation were almost as intense as those previously reported, involving increased infant crying and greatly decreased play and sexual presentation. When the mothers were returned there was a transient period of increased infantile clinging and maternal cradling but totally normal mother-infant relations were rapidly effected.

2. Experiential Variables

Maternal behavior of primiparous and multiparous rhesus monkeys has been compared by Seay (1964) using two groups of four mother-infant pairs in the standard playpen situation. Very few differences in maternal behaviors were disclosed. The general nature of the results is illustrated for maternal cradling in Fig. 12; similar results were obtained

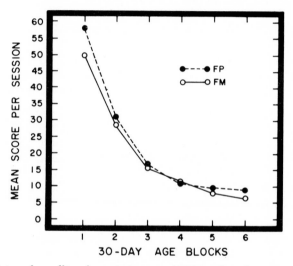

Fig. 12. Maternal cradling by primiparous (FP) and multiparous (FM) rhesus monkeys.

for maternal clinging. The frequency of maternal responding by primiparous and multiparous mothers was essentially identical and showed the same changes in frequency with increasing infant age.

One of the few reliable differences in maternal behaviors was the frequency of infant-rejecting behaviors and the age of the monkey infants when these expressions of maternal ambivalence took place. The experienced multiparous mothers exhibited rejecting behaviors earlier than the primiparous mothers and exhibited these rejecting behaviors more frequently. We have already expressed the belief that nonviolent maternal rejection at an appropriate infant age may be an effective and desirable maternal pattern, helping the infant to develop social relations with age-mates. Thus, with this possible exception, the data indicate that rhesus monkeys are effective mothers in tending their first-born infants. It is of course possible, even probable, that monkey mothers profit from infant-care experience, but this does not seem to be a variable of any major importance. The neonatal monkey responds so intimately and effectively to its mother that little nurturing experience is necessary. The neonatal human does not possess the neonatal monkey's efficient, complex mother-clinging mechanisms, and so monkey data may not generalize to man or even to anthropoid apes in this respect.

Female monkeys that fail to develop affection for members of their species in their first year of life are ineffective, inadequate, and brutal mothers toward their first-born offspring. We have had seven female rhesus monkeys that eventually became mothers, even though they had no monkey mothers of their own and no opportunity whatsoever to physically interact with age-mates during the first year of life. All seven infants would have died had we not intervened and fed them by hand. Five of the mothers were brutal to their babies, violently rejected them when the babies attempted maternal contact, and frequently struck their babies, kicked them, or crushed the babies against the cage floor, as illustrated in Fig. 13. The other two "motherless mothers" were primarily indifferent and one of these mothers behaved as if her infant did not exist.

Two of the seven infants did not survive, but the others repeatedly returned to their mothers, regardless of neglect or mistreatment. Gradually the maternal brutality became less violent. Frequency of maternal ventral contact by the infants of these motherless mothers and normal mothers was not reliably different during the fifth and sixth months of the infants' lives. Even so, the motherless mothers were still punishing their babies more violently than the normal mothers when the 6-month experiment ended. Three of these motherless mothers have delivered second babies and in all three cases they were either normal or over-protective mothers. These experiments and various interpretations have been described in greater detail by Seay et al. (1964). Thus, early social experience is an important variable for monkey mothering, but specific infant care is not an essential variable.

3. Hormonal Variables

No one can question that maternal behavior is influenced by hormonal variables. Nursing is obviously a variable influencing maternal behavior and it is obviously subject in part to endocrine control. Ovarian hormones probably influence maternal behavior, but their exact role in monkey maternal behavior has not been defined.

The strong maternal behavior of preadolescent females contrasts sharply with the relative indifference to infants shown by preadoles-

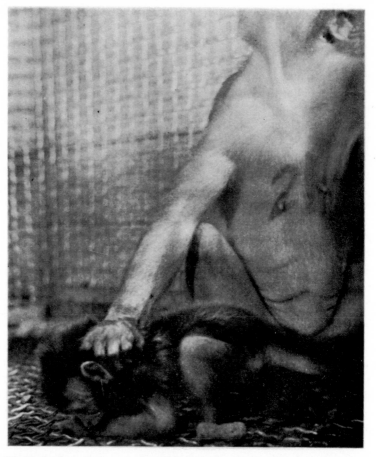

Fig. 13. Motherless mother crushing 20-day-old infant to the floor. (From Harlow, 1962c.)

cent males. The underlying variables, hormonal, neural, or both, are unknown.

Feminine Psychology

No one really understands
The female's head or heart or glands.
Perhaps it's just as well for us
That they remain mysterious.

IV. THE AGE-MATE OR PEER AFFECTIONAL SYSTEM

A third affectional system, one which we believe is vastly important to normal adolescent and adult social behavior, first develops between infants and continues to elaborate through childhood and adolescence.

This affectional system has been relatively neglected in the study of both human and nonhuman animals, and its full factual and theoretical importance has not been appreciated. The age-mate or peer affectional system doubtless continues to operate even after monkeys have become adults and then takes form primarily as affection between adult females, between adult males, and between specific heterosexual pairs. However, since this adult affectional system appears early in life, we have sometimes described it as an infant-infant affectional system, and as a child-child affectional system, as well as the age-mate or peer affectional system.

A. Development

Like all of the affectional systems, the peer affectional system goes through a series of developmental stages, each stage dependent on maturation and influenced by learning as soon as any maturational change makes its appearance. The peer affectional system breaks down into (1) a reflex stage, (2) an exploration stage, (3) a stage of interactive play, and (4) a stage of aggressive play and developing social status.

1. Reflex Stage

The reflex stage is illustrated by the behavior of infants under 20 to 30 days of age, as observed in our Playroom A, which is 8 feet high with 42 square feet of floor space and contains many stationary and mobile toys and tools, flying rings, a rotating wheel, an artificial tree, a wire-mesh climbing ramp, and a high, wide ledge, offering opportunities to explore and play in a three-dimensional world. There is a strong tendency for these infants to move about this enclosure in close physical proximity with each other, as shown in Fig. 14. If one monkey leaves our conventional group of four, either the other monkeys follow or the animal that has left returns to the group. We have hypothesized that this relatively close group-contactual relationship is a function of the visual investigatory and orienting reflexes described by Pavlov (1927, p. 12). Early in life a moving animal object elicits visual orienting responses in other members of the social group and this type of stimulation elicits the postural adjustments described by Magnus (see Evans, 1930, pp. 340–371). Thus, even in the earliest stage the infants act as an integrated social group though engaged in little behavior other than locomoting within the experimental situation and adjusting physically as best they can to the physical objects which they encounter. Reflex exploration appears during this stage and gradually declines as the infants evermore engage in voluntary exploration of all of the paraphernalia which the test situation affords.

Fig. 14. Responsiveness of 20-day-old infants in Playroom A.

2. Exploration Stage

Early exploration consists of brief sessions of gross bodily contact and oral and manual manipulation of all inanimate and animate objects present. There is limited evidence that this type of object exploration precedes social exploration. Even though object exploration becomes progressively more complex, it tends to be superseded by social exploration and the more interactive animate play patterns that will subsequently appear. The variables that elicit object exploration and social exploration are doubtless similar in kind, although they have never been subjected to detailed experimental investigation. Mobile physical objects afford a monkey an opportunity for interactive responsiveness, but no mobile object can give to an infant primate the enormous opportunity for stimulative feedback that can be achieved by contact with a social partner or partners. Our observations indicate that primates seek maximal interactive responsiveness until their age-mates develop aggressive behaviors producing mildly aversive stimulation (cf. Chapter 9 by Mason).

3. STAGE OF INTERACTIVE PLAY

The stage of play probably starts as individual activity involving very complex use of physical objects such as responding toys and flying rings, and leaping from object to object or platform to platform. When these individual play patterns are directed toward a nonresponding age-mate, we would describe them as contact-play initiation, and they are undoubtedly the precursors to the multiple and complex interactive play responses that subsequently appear. Interactive play takes many forms. The simplest seems to be a rough-and-tumble play pattern in which infant monkeys wrestle, roll, and sham bite each other without injury and seldom become frightened. We have described a second type of play which we believe to be more complex. This is approach-withdrawal play, or noncontact play, in which the monkeys chase each other back and forth with physical contact minimized. Often three or more animals engage in these noncontact play sessions, in which animal A chases animal B and/or animal C and then the role of the chaser and the chased is suddenly reversed, perhaps when the chased animal gains a physical position, such as an elevated platform, which gives him maximal opportunity to reverse the approach and withdrawal roles.

Although we believe that approach-withdrawal play is a more complex pattern than simple rough-and-tumble play, these patterns develop in parallel fashion and both increase in frequency with increasing age (Fig. 15). Furthermore, the problem of complexity is likely confounded by the tendency for males to develop a greater preference for rough-and-tumble play than females, and for females to engage in playful response patterns which are more characteristic of noncontact play. There is a progressive tendency after the sixth to eighth month for the tempo of the play patterns to increase, with frequent intermixing of contact and noncontact play to form a pattern which has been described as integrated or mixed play (Hansen, 1962; Rosenblum, 1961).

4. STAGE OF AGGRESSIVE PLAY

As the first year of life comes to a close, or the second year of life comes into being, a new play pattern called aggressive play gradually emerges. Genteel rough-and-tumble play becomes superseded by a kind of physical contact which carries with it aversive components. The animals wrestle and roll and bite as in rough-and-tumble play, but in aggressive play the manual contact and release can be physically painful and the biting responses may evoke cries of distress and anguish from the monkey being bitten. Normal rhesus monkeys are seldom permanently injured during these aggressive play encounters, but firm peer relations establishing dominance position and social status develop.

Many status relations are formed, primarily among the young males and among the young females. Once formed, they tend to remain stable for long periods of time. The status relation between males and females is more complex and the variables operating are described when we discuss the heterosexual affectional system (Section V). Somewhat similar differences in sex role in play and its relation to social patterning have been described by DeVore and Washburn (1963, p. 347).

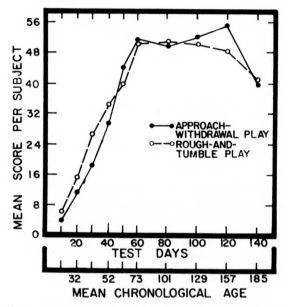

FIG. 15. Parallel development of approach-withdrawal and rough-and-tumble play.

B. Variables Determining Peer Affection

In spite of the great importance of normal peer affection in the monkey's social life, his and her heterosexual behavior, and her maternal responsiveness, we know little about the variables operating. We must accept the fact that play is no more mysterious than many other behaviors such as those of locomotion, manipulation, and exploration, and that play invariably develops in infant primates if any effective opportunity for it exists, just as locomotion and manipulation normally develop if the environment permits such behaviors.

1. EXTERNAL STIMULUS VARIABLES

Play obviously depends on visual, auditory, and proprioceptive stimulation and feedback, but the precise role of stimulation and feedback in play is yet to be defined. We are certain that the variables most important for the development of the infant-mother and mother-infant affectional systems are not the variables of primary importance for the development of the age-mate or peer affectional system.

Interanimal clinging is a primary variable underlying the infant-mother affectional system and, as we have shown, is a variable of great importance to the maternal affectional system. Contrariwise, intimate and prolonged neonatal and infantile clinging hinders the normal development of the age-mate system.

If pairs of rhesus monkeys are placed together from birth onward in a standard living cage, they rapidly develop a close clinging pattern that we have described as the two-together-together monkeys (Fig. 16). With

Fig. 16. Persisting clinging by 40-day-old two-together-together monkeys. (From Harlow, 1962a.)

the passage of time, one monkey becomes the clinger and the other the clung, so that the clung-to monkey, in fact, has a "monkey on his back." If groups of four infant monkeys are placed together in a larger empty living cage, they tend to develop a pattern of dorsoventral clinging (Fig. 17) described by us as the "choo-choo" pattern or the four-together-together monkeys. Both the two-together-together monkeys and the four-together-together monkeys showed delayed play patterns. Even

FIG. 17. Dorsoventral clinging by 7-month-old rhesus monkeys.

by the end of the first year of life, their play tended to be a slow-motion, frequently interrupted, caricature of normal infant-monkey play. When we watched the delayed development of play in these together-together monkeys, we assumed that this would have an extremely adverse effect on their heterosexual development and peer-relation development, but recent data have left this problem open. All or almost all of our together-together monkeys have developed normal patterns of adult heterosexuality and have formed relatively normal and stable peer relations. It is possible that in the protected situations of the laboratory, normal age-mate affectional adjustment may take place with a minimum of playful interaction and without maternal guidance. But the data that we have at present must be taken as being only exploratory and suggestive.

2. EXPERIENTIAL VARIABLES

Social experience during the first 6 to 12 months of life influences the capabilities of monkeys to develop effective peer-affectional interactions. This was first shown by studies on rhesus monkeys raised in the social semi-isolation of bare wire cages stacked in racks so that the infants could see and hear but not touch other infants (Harlow & Harlow, 1962). However, equally impressive studies were recently conducted by Rowland (1964), who raised a group of four infant rhesus monkeys in total social (not sensory) isolation during the first 6 months and a

FIG. 18. Six-month-old rhesus monkey in total-social-isolation apparatus.

second group of four under the same conditions during the first 12 months of life. Then he tested the capability of these monkeys to socially interact with each other and with equal-aged, social semi-isolate monkeys in our Playroom B, which was similar to Playroom A but designed to give the infants more opportunity for easy vertical mobility and more sitting space above the playroom floor.

Figure 18 shows a 6-month-old monkey in Rowland's total-social-isolation apparatus after an opaque screen and experimenter's viewing port had been raised. The terror induced in these totally isolated animals when a new world was suddenly disclosed is illustrated in Fig. 19 by the self-clutching "autistic" pattern of a 6-month totally socially isolated infant. The profound differences in play between the 6-month semi-isolated controls and the 6-month total social isolates, when placed together as a group of four in the playroom, are shown in Fig. 20. Essentially no play appeared in the total isolates until the 24th week of playroom experience, and what little play did occur was between the members of the totally isolated pair. The normal threat response, which survives semi-isolation, is almost obliterated by 6 months of total social

Fig. 19. Self-clutching by 6-month-old rhesus monkey.

isolation (Fig. 21). Furthermore, the totally socially isolated monkeys never threatened the semi-isolated infants even though they were frequently subjected to abusive and even dangerous aggression by their semi-isolated playroom partners.

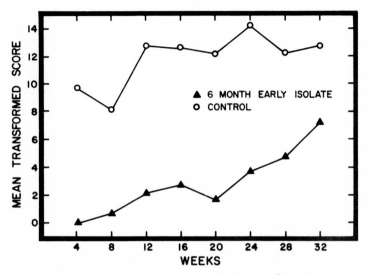

FIG. 20. Play responses by 6-month semi-isolated controls and 6-month (early) total isolates.

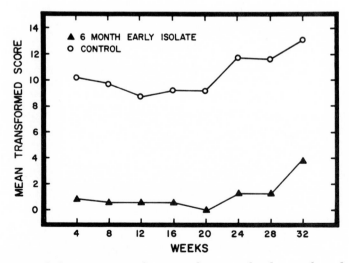

FIG. 21. Social threat responses by 6-month semi-isolated controls and 6-month (early) total isolates.

Although 6 months of total social isolation has probably irrevocably destroyed these animals' social-heterosexual capabilities (cf. Section V, B, 2), the effects of 12 months of total social isolation appear to be even more widespread. After a year of total social isolation, even the primitive oral exploratory responses have been erased from these monkeys' repertoire of behavior (Fig. 22). The effects on the elementary play pattern of contact-play initiation are equally dramatic. After the first 2 weeks, the 12-month total social isolates did not play at all, either with each other or with the pair of equal-aged semi-isolated monkeys. Essentially the same results are shown for social biting responses (Fig. 23), which develop in a relatively normal manner in the semi-isolated animals but are totally nonexistent in the 12-month totally socially deprived group.

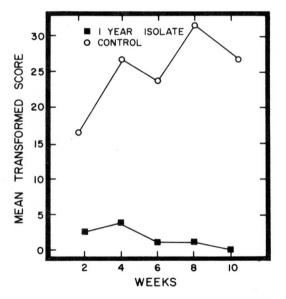

Fig. 22. Oral exploration of transportable objects by 12-month semi-isolated controls and 12-month total isolates.

On the basis of observations of the 6-month totally isolated group and their semi-isolated controls, we were convinced that a half year of total social deprivation left the isolates totally devoid of any aggressive impulses. Much to our surprise, when we paired equal-aged 6-month totally deprived and 12-month totally deprived monkeys in the playroom, the animals of the 6-month deprived group attacked their more helpless playmates in a violent and uncontrolled fashion.

In an unpublished study, G. A. Griffin has found that when rhesus monkeys are removed from 3 months of total social isolation they suffer from a brief period of shock so severe that it has been lethal in at least one case. Nevertheless, the surviving 3-month totally deprived infants have gradually adjusted to equal-aged semi-isolated monkeys, and over a period of time they apparently develop normal peer affection.

B. K. Alexander is now studying the social behavior of rhesus monkeys housed for either 4 or 8 months with their own mothers but denied any opportunity for age-mate interaction. Testing was conducted in a playpen situation in which the wire panels that separated the playpen units were replaced with opaque Masonite panels. Masonite panels were also hung between the living cages. Thus the playpen was converted into four separate compartments, each consisting of a 36-inch cubical

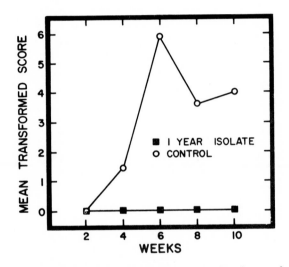

FIG. 23. Social biting responses by 12-month semi-isolated controls and 12-month total isolates.

living cage and a 30-inch cubical playpen unit joined by a 4.5-inch by 9-inch opening in the wire mesh. The monkeys in each compartment could hear the other monkeys in the room, but could not see or touch them. Infants raised only with mothers for 4 months appear to develop normal or essentially normal age-mate affectional responsiveness when given opportunity to play, but show more wariness and aggression as compared with infants raised with their real mothers but given opportunities for infant-infant interaction from 15 days of age onward. The infant monkeys raised in a mothers-only situation for 8 months also develop normal and appropriate play responses except that they are even more cautious and hyperaggressive toward each other. Such results would be consonant with the fact that animals denied the opportunity to form affectional feelings for their age-mates before aggression starts to mature are less capable of modulating and controlling these responses than normally raised rhesus monkeys.

The findings are also consistent with expectations based on mother-child relationships in the first 8 months of life. Mother monkeys, it will be recalled, are almost totally loving during the first 3 months, then show a marked increase in punitive responses. The 4-month infants had been objects of considerable monkey aggression before they began to interact with peers, and the 8-month group had met with still more aggression. Thus, these animals could view other monkeys as sources of pain and hurt as well as comfort and affection before they first encountered their age-mates. The 15-day group, on the other hand, had had primarily affectionate treatment before they associated with their peers.

Our data make it quite clear that aggression is a complex response in monkeys, modulated, inhibited, and exaggerated by various types of early social experience. Furthermore, it is evident, in keeping with Berkowitz's theory (1964), that aggression depends on an adequate aggression-evoking stimulus.

3. NEURAL VARIABLES

Evidence has ben presented by Cadell (1963) showing that bilateral section of the fornix, which doubtless produces widespread disruption throughout the rhinencephalic mechanisms, delays, damages, or destroys effective peer-affectional development. Cadell has suggested that this results from the formation of abnormal dominance relations, but the exact role of the central-nervous-system mechanisms, both neocortical and allocortical, remains to be determined.

V. THE HETEROSEXUAL AFFECTIONAL SYSTEM

The heterosexual affectional system enables adult male and female primates—prosimians, monkeys, apes, and man—to reproduce their kind.

A. Development

The heterosexual affectional system, like all of the affectional systems, goes through a series of maturational stages with the behaviors evoked in each stage being constantly subjected to learned modifications. The stages of heterosexual affectional development are (1) the reflex stage, (2) the infantile heterosexual stage, (3) the preadolescent heterosexual stage, and (4) the adult heterosexual stage.

1. Reflex Stage

The reflex heterosexual stage, perhaps better described as the reflex sexual stage, appears within a few hours or a few days after birth and is characterized by penile erection in the male and thrusting by both the male and the female. Infantile sexual thrusting may be elicited by any appropriate soft, warm object such as the body of the mother or the body of a surrogate mother. It occurs more frequently in the male infant than in the female infant, but it is relatively independent of sex. Such behavior has frequently been observed in neonatal infantile monkeys and has been reported for human infants by W. C. Lewis (personal communication). Penile erection is a response which fortunately persists for a long period of time, and reflex sexual thrusting becomes incorporated into the more complex, later maturing, heterosexual affectional system, by both the male and female, but particularly by the male.

2. Infant Heterosexual Stage

Infant sexual and heterosexual behavior is a broad and diffuse type of responsiveness which may be elicited by either animate or inanimate objects, even though animate objects—primates particularly—serve as the more adequate kind of stimulant. The incidence of the diffuse infantile sexual and heterosexual responses increases with age, but in the second half of the first year of life these responses become much more frequent and tend to change from polymorphous elicitation to relatively discrete intersex elicitation. Heterosexual responses between monkeys of the same sex, so-called "homosexual responses," persist well into adulthood, but these responses apparently have other than sexual significance—they express dominance relations and even friendships of essential social equality. It is extremely doubtful that any human type of homosexual relation exists in any nonhuman primate.

3. Preadolescent Heterosexual Stage

The appearance of the preadolescent heterosexual stage is first expressed in sex-typed response patterns, not directly related to actual heterosexual behavior. These secondary sexual behaviors occur in both the male and the female. One of the most striking of the masculine patterns is the threat response, whose developmental course in rhesus monkeys is illustrated in Fig. 24. Frequency of threat responses remains low in the female throughout the first year but progressively increases in the male during this same period. Even more striking is the social direction of these threat responses. Females may threaten females, but they seldom threaten males, whereas males threaten females and also threaten other males. There are, of course, atypical female primates, and we have seen a few females able to dominate some males in a peer group. Characteristic female secondary sexual behaviors are withdrawal responses at the time of animal interaction. The development of this pattern is traced in Fig. 25. Withdrawal is relatively infrequent in the male and relatively frequent in the female. This same sex difference holds for the postural pattern of passivity, which is characterized by im-

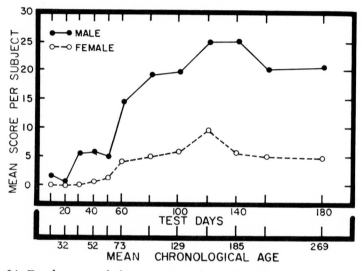

Fig. 24. Development of threat response by male and female rhesus monkeys. (From Harlow, 1962c.)

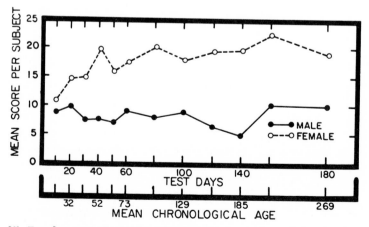

Fig. 25. Development of withdrawal responses by male and female rhesus monkeys. (From Harlow, 1962c.)

mobility with buttocks oriented toward the male and head averted, and for a similar pattern, rigidity (Fig. 26), in which the body is stiffened and fixed.

In all probability, the withdrawal and passivity behavior of the female and the forceful behavior of the male gradually lead to the development of primary sexual behaviors. The tendency of the female to orient away from the male and of the male to clasp and to tussle at the female's buttocks predisposes the consorts to assume the proper sexual positions. The development of the dorsally-oriented male sexual-behavior pattern as observed in Playroom A may be described as a composite yearning and learning curve. Early preadolescent heterosexual posturing commonly expresses itself in inadequate bodily orientation by both the male and the female (see, for example, Fig. 27, C and D). Thus, both learned

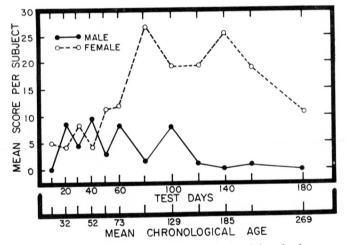

FIG. 26. Development of rigidity responses by male and female rhesus monkeys.

and unlearned variables undoubtedly operate at this preadolescent heterosexual stage to produce both heterosexual acceptance and effective heterosexual congress.

4. ADULT HETEROSEXUAL STAGE

Most monkeys eventually reach an adult heterosexual stage, but some never attain this stage of beautiful, bounteous bliss. This is true not only for monkeys that have undergone enforced social restrictions early in life in the laboratory, but it has also been reported for monkeys raised in the wild. Thus, I. DeVore (personal communication) reported the case of a very dominant adult baboon that was socially normal in every way except one—he would have nothing to do sexually with any female. Unfortunately, there is no record concerning this baboon's early social experiences but he must have met either the wrong female at the right time or the right female at the wrong time. Normal adult male and female sexual positioning is illustrated in Fig. 27, A and B, and Fig. 28. In its full-blown form it is characterized by dorsoventral mounting by

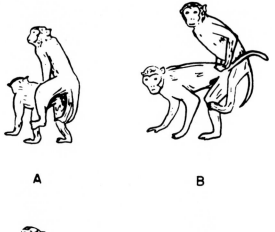

Fɪɢ. 27. A and B: Normal adult-type heterosexual responsiveness. C and D: Abnormal adult-type heterosexual responsiveness.

the male, both hands grasping the female's buttocks, and both feet clasping the female's legs. In the case of the adult male macaque, individual sexual sessions tend to be short, but it is entirely possible that the male's foot-clasp sexual pattern may be tiring. The normal female sexual positioning involves elevating the buttocks and the tail, flexing the forelimbs and shoulders, and frequently looking backward at the male either at the time of coitus or shortly afterward, no doubt with expressions of fond admiration by the normal female.

FIG. 28. Normal adult male and female sexual posturing.

B. Variables Influencing Heterosexual Affectional Behavior

1. HORMONAL VARIABLES

There can be no question that hormonal variables influence the differential nature of the heterosexual affectional system in male and female monkeys, even though most normal male and female monkeys have demonstrated an adult pattern of heterosexual behavior before the advent of gonadotropins. The appearance of these and/or the gonadal hormones no doubt affects the frequency of sexual behavior and makes possible the culmination of heterosexual congress including insemination and fertilization of the female by the male.

In an exploratory study, Goy (see Young *et al.*, 1964, pp. 214–216) obtained evidence that female monkeys, "masculinized" by repetitive injection of testosterone propionate into their pregnant mothers, tended to exhibit prepubertal masculine patterns of facial threat, play initiation, and rough-and-tumble play more frequently than normal female partners tested in the Wisconsin Playroom B situation. As the two pseudohermaphroditic females and the two normal controls were less than 6 months old, the mechanism of this effect remains to be determined. However, a hormonal variable or variables cannot be ruled out. The data present incongruities with the human hermaphroditic data presented by Hampson and Hampson (1961). Monkeys may be more delayed than human infants in the recognition of social sex role, and the social factors are undoubtedly less important for infant monkeys than for infant humans.

2. EXPERIENTIAL VARIABLES

A large body of observational data from the Wisconsin laboratories indicates that heterosexual behavior is greatly influenced by early experience, and the failure of infants to form effective infant-infant affectional relations delays or destroys adequate adult heterosexual behavior. On the basis of observational data, DeVore (personal communication) has postulated that this may be true of some olive baboons in their natural habitat.

A classic study on the effects of early experience on later heterosexual behavior was conducted by Mason (1963), who tested two groups of monkeys, half male and half female, which were somewhat over 2 years old at the time of testing. He found that none of the monkeys raised alone in wire cages for the first 6 to 12 months showed the normal pattern of heterosexual behavior. Both the males and the females were totally inept (Fig. 27, C and D). On the other hand, rhesus monkeys captured in the wild, which had certainly enjoyed normal monkey mothering and probably had ample opportunity in the first half year of life to form infant-infant affectional relations, showed perfectly normal adult-type heterosexual responsiveness (Fig. 27, A and B). Extensive studies conducted by M. G. Senko and laboratory breeding tests made by S. G. Eisele (personal communication) indicated that both adult males and females were at first totally inept in heterosexual responsiveness, even when the laboratory-raised monkeys had been placed with feral-raised males and females of our breeding colony. Even the long-term prognosis for the laboratory-raised male monkeys was very unfavorable. Out of approximately 100 males, only one male ever showed what might be called normal heterosexual behavior, and he showed this behavior only after prolonged social-sexual contacts with breeding-stock females.

Our records with female monkeys raised in social semi-isolation during the first year of life have been more hopeful. After repetitive pairings with breeding-stock males, the percentage of females that eventually accepted the male as a sexual partner increased progressively. About 50% of these females have now been inseminated and seven have given birth to baby monkeys. It appears that our population of these "motherless mothers" will double or triple during the coming year. The problem of sexual acceptance by the female is much simpler than the problem of effective heterosexual behavior by the male, since the female will be inseminated by an experienced breeding male if she will accept bodily contact without undue fear and without threat, and will support the weight of the male when he mounts. Even so, the females denied infant-infant affectional experience have vast difficulties in establishing an adequate adult-type heterosexual role, and a considerable portion of these females will be forever denied the inalienable right of motherhood. In an effort to overcome this difficulty of our reluctant females, we devised an apparatus called the rape rack. Females artificially postured in this apparatus are acceptable sexual objects to some of our breeding males and not acceptable sexual objects to others. This fact should cause no surprise in view of a vast wealth of human clinical data. The little data we now possess suggests that rape-rack experience does not overcome female frigidity—a finding not completely surprising.

G. W. Meier, studying the behavior of laboratory-raised rhesus monkeys at the University of Puerto Rico, has reported normal sexual behavior by both males and females that have been raised from birth onward in bare laboratory cages, but denied opportunity early in life to interact effectively with other infants. He has suggested that the difference lies in the caging conditions, believing that his cages were more closely intermeshed, even though physical contact between infants was denied. Perhaps normal sexual behavior might develop despite inadequate infant-infant affectional development if the infants could observe adult sexual behavior. However, Gene Sackett, a member of the Wisconsin Primate Laboratory staff, visited the Puerto Rican Laboratory and reported that Dr. Meier believes the Puerto Rican monkeys did not view adult sexual behavior early in life. As neither Dr. Meier nor Dr. Sackett could suggest any definitive answers at this time, the variables that produce the difference between the behaviors of the Wisconsin and Puerto Rican monkeys can only be sought by further experimentation.

We have monkeys that have been raised in total social isolation without visual, auditory, or bodily contact with other animals after being separated from their mothers a few hours after birth (Section IV, B, 2). Because of their severe social impairment, we seriously doubt that any of these 6- and 12-month totally isolated animals will ever develop a normal adult-type heterosexual behavioral system, although this conjecture must await the test of time. Meanwhile, we have observed that prolonged social semi-isolation or 6 months of total social deprivation does not deny the rhesus monkey the privilege and pleasure of auto-erotic, masturbatory responses. But even this infantile form of sexuality is either deeply depressed or completely eliminated after 12 months of total social isolation.

At present, we must conclude that the data on early experience in relation to adult heterosexuality in monkeys afford nothing more than hot and cold leads, but that a large part of the data affords leads approaching frigidity.

3. NEURAL VARIABLES

Electrical stimulation of subcortical nuclei in the dorsal hypothalamus and probably the midbrain will elicit penile erection in the male squirrel monkey (*Saimiri sciureus*) (MacLean & Ploog, 1962). These data, beautiful as they are, still leave us up in the air concerning the neural variables that integrate the total pattern of heterosexual affectional behavior in either male or female monkeys. Klüver and Bartelmez (1951) reported one case of a female rhesus monkey that behaved very much like our motherless monkeys following radical bilateral lobotomy of the temporal lobes, and it is possible that the centers buried within the medial and rostral temporal-lobe masses may play an important role in organizing female heterosexuality. A similar observation was made by Walker *et al.* (1953). Delgado's (1963, 1964) techniques of telemetering electrical stimulation to subcortical nuclei in restricted and nonrestricted rhesus monkeys offer hopeful leads toward the eventual resolution of many neurological problems related to the operation of the heterosexual affectional system, but these studies are still at an exploratory level.

VI. THE PATERNAL AFFECTIONAL SYSTEM

The primate paternal affectional system expresses itself in the behavior of the adult male to the members of his social group, whether this group is a monogamous family, polygamous family, clan, horde, or tribe. The paternal affectional system involves care and protection of the female and her offspring, and frequently intimate relationships with the male's own children or substitute children. It has long been known that man is an atypical primate anatomically, physiologically, and psychologically, and the paternal affection exhibited by adult dominant nonhuman males under feral conditions differs from that of the family-plan human parent. Male monkeys in the wild should be regarded as generalized fathers: they show affectional responses to members of their social group but do not show them differentially to their own or other children. The dominant male, or dominant males when such exist in fairly large groups, protect all members of their social group against aggression at the risk of their own lives. Field studies of baboons indicate that group-protective functions may be served by nondominant and young adult males as well as dominant males, and the social protective functions of these males differ (DeVore & Washburn, 1963, pp. 343–345).

Dominant male monkeys also maintain a social order within the group and usually will not permit a large infant to abuse a small infant in the absence of the small infant's mother. There are authenticated cases of large dominant males adopting for a considerable time a specific youngster, usually a youngster of about 1 year of age, guarding it against all other members of the monkey group and even carrying the young on their bodies, either on the ventral or the dorsal surface. This has been reported for free-ranging olive baboons by DeVore (1963), for a number of groups of Japanese macaques (*Macaca fuscata*) by Itani (1963; see also Chapter 15 by Jay), and for bonnet macaques in the laboratory (Fig. 29) by I. C. Kaufman and L. A. Rosenblum (personal communica-

Fig. 29. Male bonnet macaque holding infant.

tion). Threat to an infant monkey may produce violent aggressive responses by adult males; an illustrative laboratory example has been given by Bernstein (1964). Probably the most striking example is the case of a cowardly, dominant male bonnet macaque, as reported by L. A. Rosenblum (personal communication). The male, one or more adult females, some juveniles, and offspring lived communally in the family-plan laboratory situation illustrated in Fig. 30. The male was terrified when any adult human being entered the monkey living quarters

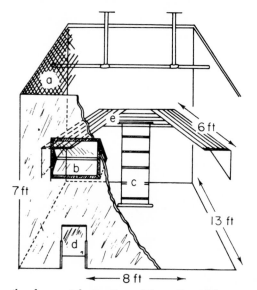

Fɪɢ. 30. Family-plan social situation. (From Rosenblum *et al.*, 1964.)

and would retreat to a corner and scream in fright. On one occasion, however, the door to the living quarters was left open and an infant wandered out. Without warning, the cowardly male attacked Dr. Rosenblum, who was finally able to escape and lock the cage door leaving the cowardly male inside and himself outside.

The complex operations of the paternal affectional system in the Japanese macaque have been detailed by Itani (1963). Analysis of the development of the paternal affectional system under laboratory conditions and determination of its underlying variables are yet to be achieved. Such information might be obtainable in the "family plan" laboratory facilities of Rowell and Hinde (1963) and Kaufman and Rosenblum, and some of the laboratory-raised male monkeys at the Wisconsin Primate Laboratory are apparently approaching a relatively normal male maturity. For the present time, however, the nonhuman primate parent is best described in the many exemplary field studies.

VII. SUMMARY

We are convinced that there are at least five affectional systems in the primate order: the systems of infant-mother affection, mother-infant affection, peer affection, heterosexual affection in adults, and paternal affection. We believe these systems all go through an orderly series of maturational stages and we also believe that they operate through different behavioral, neural, and biochemical variables. Some of the systems, such as the mother-infant and the infant-mother system, depend on similar variables, but in other systems, such as the age-mate or peer affectional system, the variables differ strikingly from those found elsewhere.

REFERENCES

Berkowitz, L. (1964). Aggressive cues in aggressive behavior and hostility catharsis. *Psychol. Rev.* **61**, 104.

Bernstein, I. S. (1964). Role of the dominant male rhesus monkey in response to external challenges to the group. *J. comp. physiol. Psychol.* **57**, 404.

Cadell, T. E. (1963). The effects of fornix section on learned and social behavior in rhesus monkeys. Doctoral dissertation, University of Wisconsin. University Microfilms, Ann Arbor, Michigan, No. 63-7568.

Delgado, J. M. R. (1963). Telemetry and telestimulation of the brain. *In* "Bio-Telemetry" (L. Slater, ed.), pp. 231–249. Pergamon Press, New York.

Delgado, J. M. R. (1964). Free behavior and brain stimulation. *Int. Rev. Neurobiol.* **6**, 349.

DeVore, I. (1963). Mother-infant relations in free-ranging baboons. *In* "Maternal Behavior in Mammals" (Harriet L. Rheingold, ed.), pp. 305–335. Wiley, New York.

DeVore, I., & Washburn, S. L. (1963). Baboon ecology and human evolution. *In* "African Ecology and Human Evolution" (F. C. Howell & F. Bourlière, eds.), pp. 335–367. Wenner-Gren Foundation for Anthropological Research, New York.

Evans, C. L. (1930). "Recent Advances in Physiology," 4th ed. P. Blakiston's Son, Philadelphia, Pennsylvania.

Hampson, J. L., & Hampson, Joan G. (1961). The ontogenesis of sexual behavior in man. *In* "Sex and Internal Secretions" (W. C. Young, ed.), 3rd ed., Vol. II, pp. 1401–1432. Williams & Wilkins, Baltimore, Maryland.

Hansen, E. W. (1962). The development of maternal and infant behavior in the rhesus monkey. Doctoral dissertation, University of Wisconsin. University Microfilms, Ann Arbor, Michigan, No. 63-653.

Harlow, H. F. (1958). The nature of love. *Amer. Psychologist* **13**, 673.

Harlow, H. F. (1959). Love in infant monkeys. *Sci. Amer.* **200** (6), 68.

Harlow, H. F. (1962a). Development of the second and third affectional systems in macaque monkeys. *In* "Research Approaches to Psychiatric Problems" (T. T. Tourlentes, S. L. Pollack, & H. E. Himwich, eds.), pp. 209–229. Grune & Stratton, New York.

Harlow, H. F. (1962b). The development of affectional patterns in infant monkeys. *In* "Determinants of Infant Behaviour" (B. M. Foss, ed.), pp. 75–88. Wiley, New York.

Harlow, H. F. (1962c). The heterosexual affectional system in monkeys. *Amer. Psychologist* **17**, 1.

Harlow, H. F., & Harlow, Margaret K. (1962). Social deprivation in monkeys. *Sci. Amer.* **207** (5), 136.

Harlow, H. F., & Zimmermann, R. R. (1959). Affectional responses in the infant monkey. *Science* **130**, 421.

Harlow, H. F., Harlow, Margaret K. & Hansen, E. W. (1963). The maternal affectional system of rhesus monkeys. *In* "Maternal Behavior in Mammals" (Harriet L. Rheingold, ed.), pp. 254–281. Wiley, New York.

Imanishi, K. (1957). Social behavior in Japanese monkeys, *Macaca fuscata. Psychologia* **1**, 47.

Itani, J. (1963). Paternal care in the wild Japanese monkey, *Macaca fuscata. In* "Primate Social Behavior" (C. H. Southwick, ed.), pp. 91–97. Van Nostrand, Princeton, New Jersey.

Jay, Phyllis (1963). Mother-infant relations in langurs. *In* "Maternal Behavior in Mammals" (Harriet L. Rheingold, ed.), pp. 282–304. Wiley, New York.

Klüver, H., & Bartelmez, G. W. (1951). Endometriosis in a rhesus monkey. *Surg., Gynecol. Obstet.* **92**, 650.

MacLean, P. D., & Ploog, D. W. (1962). Cerebral representation of penile erection. *J. Neurophysiol.* **25**, 29.

Mason, W. A. (1963). The effects of environmental restriction on the social development of rhesus monkeys. *In* "Primate Social Behavior" (C. H. Southwick, ed.), pp. 161–173. Van Nostrand, Princeton, New Jersey.

Mowbray, J. B., & Cadell, T. E. (1962). Early behavior patterns in rhesus monkeys. *J. comp. physiol. Psychol.* **55**, 350.

Pavlov, I. P. (1927). "Conditioned Reflexes." Oxford Univer. Press, London.

Rosenblum, L. A. (1961). The development of social behavior in the rhesus monkey. Doctoral dissertation, University of Wisconsin. University Microfilms, Ann Arbor, Michigan, No. 61-3158.

Rosenblum, L. A., Kaufman, I. C., & Stynes, A. J. (1964). Individual distance in two species of macaque. *Anim. Behav.* **12**, 338.

Rowell, T. E., & Hinde, R. A. (1963). Responses of rhesus monkeys to mildly stressful situations. *Anim. Behav.* **11**, 235.

Rowland, G. L. (1964). The effects of total social isolation upon learning and social behavior of rhesus monkeys. Doctoral dissertation, University of Wisconsin. University Microfilms, Ann Arbor, Michgian, No. 64-9690.

Seay, B. M. (1964). Maternal behavior in primiparous and multiparous rhesus monkeys. Doctoral dissertation, University of Wisconsin. University Microfilms, Ann Arbor, Michigan, No. 64-10, 305.

Seay, B., & Harlow, H. F. (in press). Maternal separation in the rhesus monkey. *J. nerv. ment. Dis.*

Seay, B., Hansen, E., & Harlow, H. F. (1962). Mother-infant separation in monkeys. *J. child Psychol. Psychiat.* **3**, 123.

Seay, B., Alexander, B. K., & Harlow, H. F. (1964). Maternal behavior of socially deprived rhesus monkeys. *J. abnorm. soc. Psychol.* **69**, 345.

Walker, A. E., Thomson, A. F., & McQueen, J. D. (1953). Behavior and the temporal rhinencephalon in the monkey. *Bull. Johns Hopkins Hospital* **93**, 65.

Washburn, S. L., & DeVore, I. (1961). The social life of baboons. *Sci. Amer.* **204** (6), 62.

Young, W. C., Goy, R. W., & Phoenix, C. H. (1964). Hormones and sexual behavior. *Science* **143**, 212.

Chapter 9

Determinants of Social Behavior in Young Chimpanzees[1]

William A. Mason

Delta Regional Primate Research Center, Covington, Louisiana

I. INTRODUCTION

A. Chimpanzee Infancy and Childhood

The central figure in the world of the newborn chimpanzee (*Pan*) is its mother. She is the source of nourishment, warmth, and protection, and the touchstone of emotional security. The infant's principal postural adjustment is clinging, which insures the close physical relation to the mother that is essential to survival in the wild. For several weeks after birth, contact with the mother is continuous. As the infant's strength and motor coordination improve, it begins to crawl about on the mother and

[1] Work on unpublished experiments described in this chapter was supported by research grants (M-4100 and M-5636) from the National Institute of Mental Health, U. S. Public Health Service. I was assisted in these studies by R. Chang-Yit, B. Hare, V. Hayhurst, N. Itoigawa, S. Saxon, and L. Sharpe.

to explore her body, much as it will later investigate the physical environment. Locomotor independence is achieved between 4 and 6 months of age, after which the amount of time spent away from the mother increases steadily. Solid foods are accepted near the middle of the first year, and at about this time appear the first clumsy attempts at social play and grooming (Jacobsen *et al.*, 1932; Tomilin & Yerkes, 1935; Yerkes & Tomilin, 1935). Infancy, characterized by extreme physical and emotional dependence on the mother, draws to a close sometime before the end of the second year of life. Six years of childhood lie ahead.

From the human point of view, chimpanzee childhood is an altogether delightful time of life. It has been described as a period of ". . . independence and freedom for play, with a minimum of social responsibility and personal risk. The hampering limitations of infancy have been escaped—the individual is no longer tied to the mother's apron strings—and the major burdens of self-maintenance and defense, of reproduction and social cooperation, have not yet been assumed" (Yerkes, 1943, p. 55).

Standing as it does between the helplessness of infancy and the self-sufficiency of adulthood, chimpanzee childhood presents an intriguing spectacle to the psychologist. Childhood is, conspicuously, a period of intense sociability, characterized by frequent and unmistakable displays of affection, playfulness, and social dependence. One has only to observe a pair of compatible young chimpanzees that have lived together for a short time to be convinced of the central place that each animal occupies in the life of the other. One is also impressed with the complexity of their relationship and with the frequent and sometimes mercurial changes in behavior toward one another.

B. The Present Approach

The determinants of such moment-to-moment variations in the social activity of young chimpanzees is the principal concern of this chapter. Its purpose is to describe the normal patterns of social interaction between young animals, to illustrate how variations in these patterns may be induced, and to suggest a mechanism that may help to account for such variations. The thesis to be developed in the following sections revolves around two generalizations: First, a given motivational state, which may be thought of as level of arousal (Bindra, 1959), predisposes the animal to engage in characteristic patterns of social activity. Second, these activities themselves influence the existing level, either raising it or lowering it. Thus, a homeostatic relation between level of arousal and social behavior is possible. If variations in arousal level produced in this manner are reinforcing, a mechanism is suggested through which the performance of particular social responses may become habitually associated with a specific companion, thus giving rise to the familiar primate social phenomena of "friendships," "attachments," "emotional dependence," and "companionship preferences."

C. Social Responses

For some purposes, social responses may be defined as those that are elicited by social stimuli and have some effect on other individuals (Nissen, 1951). Our interest, however, will center on a few relatively stereotyped (species-typical) behavior patterns which can be described

Fig. 1. Clinging. (Yerkes Laboratories photograph.)

in terms of their *form*. We will be particularly concerned with three such patterns: *clinging, grooming,* and *play-fighting.* These patterns are most fully and characteristically expressed in social intercourse, but they can and do occur with inanimate objects. Each pattern can be described and classified in terms of the spatiotemporal organization of its component responses and without reference to eliciting stimuli or goal objects. A brief description of each pattern follows:

1. Clinging. Clinging characteristically includes grasping and ventral contact with a social stimulus (Fig. 1), and only this pattern was scored as clinging in our research. Other behaviors such as tandem walking and draping an arm around a companion (Bingham, 1928; Schiller, 1957) are probably variants of clinging. Chimpanzees sometimes cling to inanimate objects (McCulloch, 1939; see also below), and the occasional posture of young chimpanzees, sitting with the arms folded across the chest while the hands grasp a contralateral limb (Fig. 2), may be a self-directed form of this pattern.

2. Grooming. Grooming may range from rather casual picking and manipulation of the body surface of a companion to an intense activity which includes close visual regard, coordinated movements of the fingers, and lip-smacking or sputtering (Fig. 3; Yerkes, 1933). Components of the grooming pattern may occur with nonsocial stimuli (Schiller, 1952) or grooming may be self-directed. Since young animals (less than 4 years) rarely show the complete grooming pattern (Nowlis, 1941; Yerkes, 1933), mild or incomplete forms of grooming have been scored in the present research.

3. Play-fighting. Play is the most complex and variable of the three patterns. It consists of grappling, rolling, slapping, pulling, pushing, and mouthing or mock-biting (Fig. 4). Play is often accompanied by a characteristic open-mouthed expression ("smiling") and by panting vocalizations. This pattern has been observed to occur with inanimate objects (Fig. 5; Bernstein, 1962; Menzel, 1963; Schiller, 1957) and is occasionally seen in the absence of an apparent eliciting stimulus, as a kind of vacuum reaction. Play-wrestling is frequently interwoven with noncontact forms of play such as chasing or "tag." In the research described here only contact play was scored.

In most of our experiments, the method used to quantify social behavior has been the time-ruled checklist, with behavior recorded by 15-second intervals. Thus, in this chapter, unless otherwise noted, reference to the occurrence of a response should be taken to mean the occurrence of one or more such responses in a 15-second interval. The reliability of this method, determined by comparing the performance of two persons independently scoring the same behavior episode, generally ranges between 70 and 95%.

Fig. 2. "Huddling." Possibly a self-directed form of clinging. (Yerkes Laboratories photograph.)

Fɪɢ. 3. Grooming. (Yerkes. Laboratories photograph.)

FIG. 4. Social play (play-fighting). (Yerkes Laboratories photograph.)

In some experiments, we have used clinging, grooming, and play as independent variables. In such cases the patterns have been simulated by a human (stimulus-person). In many experiments, stimulus-persons have worn distinctive costumes, including simple masks, which concealed the identity of the person and provided a visual indication of the role. Further, to minimize the effects of individual differences in role playing, stimulus-persons have systematically interchanged roles within each experiment.

II. SOCIAL BEHAVIOR OF YOUNG CHIMPANZEES: NORMATIVE ASPECTS

A. Interactions of Wild-Born Chimpanzees

To provide a description of the frequency and form of social interaction occurring between young animals in an ordinary laboratory setting, seven pairs of wild-born chimpanzees about 2 to 4 years old were observed for twenty 30-minute periods in a familiar outdoor

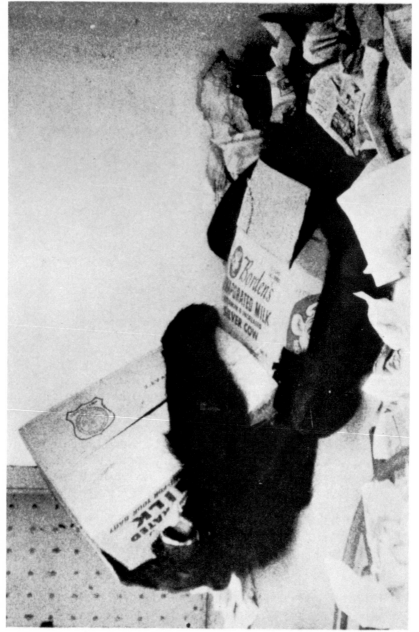

Fig. 5. Play-fighting with box. (From Bernstein, 1962.)

cage. Clinging, grooming, and play were the most frequent social activities, together accounting for over 79% of all specific social contacts. Clinging accounted for 4% of this total, grooming for 5%, and play for 70%. What accounts for this hierarchy in the frequency of social activities?

B. Social Activities as Rewards

It might be expected that the different frequencies of social responses observed during normal interaction reflect differences in the effectiveness of these activities as rewards (Falk, 1958). To test this hypothesis five preadolescent laboratory-born chimpanzees were given the opportunity to engage in four social activities with a familiar stimulus-person: He (1) played with them, (2) petted them, (3) groomed them, or

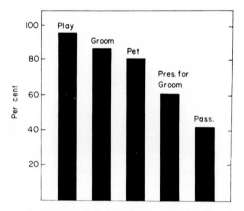

FIG. 6. Percentage of 15-second intervals in which the chimpanzees were in proximity to stimulus-person. (After Mason *et al.*, 1962.)

(4) allowed them to groom him (Mason *et al.*, 1962). The animals were tested in their living cages, and the person remained outside the cage and interacted with the subjects through the 2.5-inch wire fencing. Each test session consisted of the presentation of one of the four social activities for 150 seconds, preceded and followed by periods of the same duration during which the stimulus-person remained passive. Data were obtained on a variety of measures (see Section IV, C), but the most direct indication of reward effects was the percentage of 15-second intervals in which the chimpanzee was at the front of the cage and directly opposite the stimulus-person. These proximity scores (Fig. 6) were highest for play (95.8%) and lowest for present-for-grooming (61.2%) among the active conditions. For purposes of comparison, scores for the first passive condition are also shown.

A second experiment, also reported by Mason *et al.* (1962), sought to replicate these findings using younger animals tested by a different method and in a different situation. The subjects were two African-born chimpanzees, about 3 years old. The apparatus consisted of two interconnected compartments, each equipped with a lever-operated retractable window. Again, social stimulation was provided by a human companion, but instead of the single-stimulus method of the first experiment, a paired-comparison technique was used and the social stimulation was contingent on lever pressing. Social activity preferences (Fig. 7) were similar to those obtained in the first study.

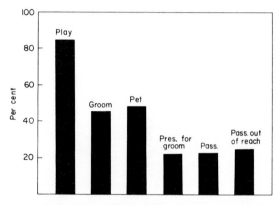

Fig. 7. Percentage of trials on which different forms of social stimulation were chosen in paired-comparison tests. "Pass. out of reach" refers to a condition in which the stimulus-person remained both passive and out of reach of the animal. (After Mason *et al.*, 1962.)

As a final check on the stability of these effects and to provide a comparison between social rewards and more conventional reinforcers, two additional experiments were performed. The dual-cage apparatus was again used with a paired-comparison method, and three African-born animals were subjects. In the first of these experiments, the rewards used during each test session were food and social stimulation. As social rewards we selected play, the most preferred activity in the foregoing studies, and petting, which consistently occupied an intermediate position. The animals were tested just after their regular morning feeding, or they were tested immediately before the morning feeding when they had been deprived of food for at least 15 hours. Figure 8 shows the percentage of choices of social rewards when compared with food. The relative differences between play and petting agree with previous findings, and it can be seen that, when the animals had been recently fed, play was actually preferred over food. In the next experiment, play and petting were again compared with food rewards, but in this case the foods had high, moderate, or low preference values. Again, there was a consistent difference between play and petting, and the preference for these activities was inversely related to the preference value of the food.

C. Development of Social Activity Preferences

We are not sure at what point in individual development the preference for play first appears. Almost certainly it is not present during the first few weeks of life, but it appears some time before the end of the

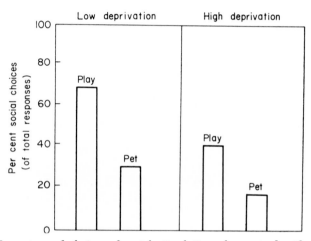

FIG. 8. Percentage of choices of social stimulation when paired with food, under two levels of food deprivation.

first year. In one experiment, five infant chimpanzees were given two 20-minute "conditioning" periods each day starting at the age of 126 days. During each of these periods, they were placed in a room with a stimulus-person dressed in one of two distinctive costumes. The role associated with one costume consisted of holding the chimpanzee in a ventroventral position (Fig. 9); for the other costume the role was play (tickling, pulling, and pushing). At the end of each conditioning period the animal was placed at one side of the room while the stimulus-person seated himself at the opposite side, and a record was obtained of the percentage of 15-second intervals during which the animal contacted the person during a 5-minute period. No evidence of preference was found in the first 40 test sessions, but interpretation of this outcome is complicated by the fact that during the first few months of testing some subjects

FIG. 9. Costume associated with holding.

were still not locomoting efficiently. The preference for play was first evident in the third block of 20 sessions, and it persisted throughout the remainder of the experiment (Fig. 10).

In general, the findings of the entire series of experiments are quite consistent. Play is the activity of choice for young chimpanzees, and this is true whether preference is measured by responses to a human companion or inferred from the normal interaction between cagemates. The preference is present before the end of the first year of life, persists until at least early puberty, and, within the limits of our subject population, is independent of conditions of rearing. In each of the foregoing experiments, however, care was taken to insure that the animals were well adjusted to the situation and to their companions before the data were obtained. We shall see that this is an essential condition for the effects I have described.

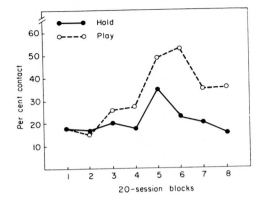

FIG. 10. Development of preferential responses to play and holding.

III. EFFECTS OF AROUSAL LEVEL ON SOCIAL RESPONSIVENESS

A. Situational Factors

In his studies of claspable objects as rewards for delayed response and discrimination learning by young chimpanzees, McCulloch (1939) noted that such objects were maximally effective only when the subjects appeared excited or disturbed. This may be interpreted as indicating that the tendency to cling increases with increasing arousal. In one test of this hypothesis we used three young laboratory-born chimpanzees, raised from infancy with cloth artificial mothers similar to those used by Harlow and Zimmermann (1958). The chimpanzees' responses to the artificial mother and to a passive person with whom they had had frequent but limited contacts were compared in seven situations designed to produce different degrees of arousal: (A) Both stimulus objects were present and accessible. (B) One of the stimulus objects was screened, so that it could be seen but not contacted. (C) Only one stimulus object was present. (D) Same as C, with 1 minute of restraint before testing. (E) Same as C, with intermittent noise during testing. (F) Same as C, with novel objects present. (G) Same as C, but out-of-doors. Tests A through F were conducted in a familiar observation room. Every animal received twenty 5-minute tests in each situation. Figure 11 shows the percentage of 15-second intervals in which the chimpanzees contacted each stimulus object. In situations A and B, in which arousal was presumably low, the person received more contacts than did the surrogate, and many of these responses were playful or investigatory. In tests C through G, which were designed to produce progressively higher degrees of arousal, contacts with the artificial mother increased progressively, whereas the reverse trend was observed for the person. Thus, the results indicated a tendency for contacts with an object habitually associated with clinging to increase as a function of increasing arousal. Under the same conditions, contacts with a less familiar stimulus object which seemed to evoke principally play and investigatory activities tended to decrease, even when it was the only stimulus object available.

Perhaps placing the chimpanzees in a distressing situation merely caused them to seek the more familiar object and to avoid the less famil-

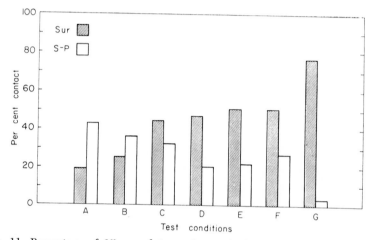

Fig. 11. Percentage of 15-second intervals in which chimpanzees contacted artificial mother (Sur) and stimulus-person (S-P).

iar, in which case more specific functions associated with these objects need not be invoked. Other experiments suggest that this is unlikely. In one experiment, 12 African-born chimpanzees were given a choice between two costumed stimulus-persons. As in the study mentioned in Section II, C, one of the stimulus-persons responded by initiating play when the animal approached, and the other, by drawing the subject to him and holding it in a ventroventral position. The animals had had equal previous exposure to the two social stimuli. Tests were conducted in a familiar room and in one which was unfamiliar. Contacts with the stimulus-person who held the subject are presented in Fig. 12 as a percentage of contacts with both stimulus-persons. In the familiar situation there was a clear preference for play, while in unfamiliar surroundings holding was preferred. With repeated exposures to an unfamiliar situation habituation occurred, and a consistent preference for play appeared within eight sessions.

In addition to methods that require the animal to choose between stimuli representing different kinds of social activity, one may demonstrate the contrasting relations of play and clinging to arousal level by measuring changes in the *form* of the response to the same stimulus presented in different situations. In one experiment, for example, five pairs of wild-born chimpanzees were observed in three situations differing in degree of familiarity. Each pair was observed for three 5-minute sessions in its living cage (home cage), and in an identical cage placed in a different location in the colony room (holding cage) or in an unfamiliar room. Clinging responses as a percentage of total clinging and play re-

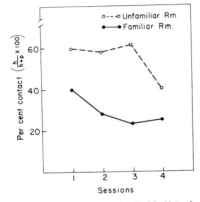

FIG. 12. Contacts with stimulus-person who held (h) the subject as a percentage of total social contacts (h + p), in familiar and unfamiliar situations.

sponses increased progressively as a function of situational novelty. Scores in the living cage, holding cage, and unfamiliar room were 15%, 23%, and 51%, respectively. In the unfamiliar room, clinging decreased from 99% in the first session to 23% in the third. In another experiment we found that the three laboratory-born chimpanzees raised with cloth artificial mothers showed a reduction in play and an increase in the relative frequency of clinging to familiar inanimate objects (e.g., rag mop) as a function of increasing novelty of the situation. The percentage of clinging responses was 23% in the living cage, 43% in a moderately familiar cubicle, and 54% in an unfamiliar open-field situation.

The data presented thus far indicate that gross changes in the familiarity of the test situation depress play behavior and strengthen the tendency to cling. Under some circumstances, however, novelty may enhance play behavior. This seems to occur chiefly in situations in which the departure from the familiar is not extreme or to which the animal is partly habituated. Figure 13 shows the relative frequency of play for six pairs of wild-born chimpanzees. The animals were observed for 20-minute sessions while in a test cage located in the room in which they were housed and while in an identical cage placed in a situation to which they had been previously exposed only during brief experimental sessions. After the first few sessions, play was consistently more frequent in the moderately unfamiliar situation. These and similar observations suggest that increasing arousal may either augment or inhibit playfulness depending on the level achieved.

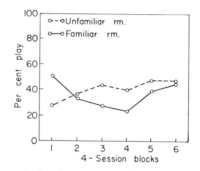

Fig. 13. Play responses, in familiar and moderately unfamiliar situations, as a percentage of total clinging and play responses.

B. Social Deprivation

Separation from companions is an acutely distressing experience to young chimpanzees (Kellogg & Kellogg, 1933; Köhler, 1927). The initial reactions include a sharp increase in screaming and whimpering and an upsurge of violent motor activity such as pacing, leaping, or shaking the cage. These responses gradually subside, but some evidence of distress may persist for many hours following separation. We have used separation of cagemates as another means of manipulating arousal level. Twelve young wild-born chimpanzees caged in pairs were observed for 20-minute periods following each of four conditions of separation: (1) no separation, (2) minimum separation, (3) 1-hour separation, and (4) 3-hour separation. Condition 1 consisted of observing interactions in the living cage in the absence of any experimental manipulation. Condition 2 provided a control for the excitement induced by separation itself; the animals were separated and immediately reunited. Figure 14 shows the contrasting relations between clinging, grooming, and play behavior. Clinging was the most frequent response initially for all conditions in which separation occurred, whether for 3 hours or merely a few seconds. It is apparent, however, that the maxima for clinging for the different separation conditions varied in orderly fashion, the highest occurring after the 3-hour separation and the lowest after no separation. Under all conditions involving separation, clinging decreased rapidly, reaching levels below 10% within the first 5 minutes. The chimpanzees played more after minimal separation than after none at all, providing further support for the thesis that moderate increments in arousal facilitate social play. Another point of interest revealed by this experiment is that grooming tends to intervene between clinging and play.

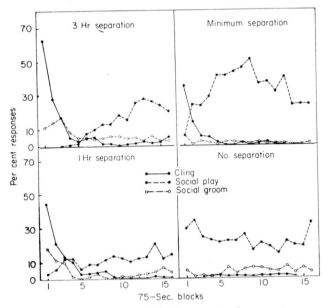

FIG. 14. Effects of social deprivation on social responsiveness.

IV. FUNCTIONS OF SOCIAL RESPONSE PATTERNS

A. Social Response Patterns as Aspects of Approach-Withdrawal

We have seen that patterns of social responsiveness can be altered by a variety of procedures designed to manipulate arousal level. In general, the findings suggest that low or moderate levels of arousal predispose the young primate to play, while high levels of arousal produce avoidance of play and strengthen the tendency to cling. In the present section, we will be concerned with the psychological functions these patterns appear to serve.

In approaching this question it will be helpful to consider play and clinging in ontogenetic perspective. Clinging is, of course, the first of these patterns to appear in ontogeny, and its role in the maintenance of bodily contact with the mother is fundamental. Play emerges somewhat later. Although a mother chimpanzee and her infant may play occasionally (Bingham, 1927; Yerkes & Tomilin, 1935), play is primarily an activity of the young (Nissen, 1931) and characteristically occurs after the infant has taken its first steps away from the mother (see Chapter 15 by Jay).

Whether or not a specific stimulus will elicit approach and play depends in part on the nature of the subject's previous experience with it and with similar objects. Investigations of young chimpanzees' responses to inanimate objects indicate that the initial reaction to a stimulus object representing a new "class" is characterized by extreme caution. With repeated exposure to comparable objects a generalized habituation occurs, and new objects of the same type are contacted freely (Menzel, 1963; Menzel et al., 1961). Moreover, when this point has been reached, the novel stimulus generally elicits more contacts than the familiar one (Menzel et al., 1961; Welker, 1956). A similar situation exists in respect to social stimuli. The home-reared chimpanzee, Viki, preferred strangers over familiar persons as playmates (Hayes & Hayes, 1951), while laboratory chimpanzees of comparable age whose contacts with humans are limited may show marked timidity in the presence of strangers, which is only gradually overcome (Hebb & Riesen, 1943; Mason et al., 1962). As habituation proceeds, however, the novel stimulus generally becomes preferred. When chimpanzees raised with artificial mothers were allowed to choose between them and devices of similar form, but differing from the original in color, they eventually came to make more contacts with the novel objects. Furthermore, play responses were directed toward the novel objects more than five times as frequently as toward the familiar one. In another experiment, described more fully in Section IV, C, chimpanzees were allowed to choose between two passive stimulus-persons, one wearing a familiar costume previously associated with play and the other wearing a novel costume, and the novel costume was preferred.

Schneirla (1959) has proposed that approach and withdrawal are fundamental dimensions of behavioral organization for all animals. Low or moderate stimulation is said to facilitate approach, while higher intensities of stimulation lead to withdrawal. A consideration of the ontogeny of social play and of the findings I have described supports the conclusion that play is essentially an approach activity, directed toward those aspects of the environment that are mildly variable or strange. I propose that clinging, on the other hand, is functionally similar in primates to physical withdrawal. In nature, seeking the mother in a disturbing situation usually results both in clinging and in physical withdrawal from the situation. As McCulloch (1939) noted, however, the act of clinging, rather than physical withdrawal, is the critical event. In fact, it is possible to arrange the situation so that the animal is actually required to move closer to the source of its distress in order to attain an object of clinging. For example, infant rhesus monkeys (*Macaca mulatta*) approached a fear-producing stimulus object in order to contact an artificial mother (Harlow, 1960). In another part of the experiment with chimpanzees raised with artificial mothers (Section III, A) we sought to compare the functional similarity of clinging and withdrawal. A stimulus-person, the artificial mother, and a small enclosed cage used to transport the subjects to and from the test area were presented in paired-comparison tests conducted in a familiar observation room and in the strange environment of an open field. Each stimulus pair was presented for twenty 5-minute sessions in both situations. The subjects preferred the person over either inanimate object when they were tested in the familiar room, whereas in the open field the preference for the cage and the cloth surrogate was unequivocal (Fig. 15). When the cage and the

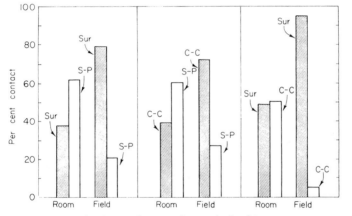

FIG. 15. Percentage of 15-second intervals in which chimpanzees contacted artificial mother (Sur), carrying cage (C-C), and stimulus-person (S-P), during paired-comparison tests in familiar observation room and unfamiliar open field.

artificial mother were compared in the open field (Fig. 15, right panel), the animals overwhelmingly preferred the surrogate even though this kept them in the situation, while the cage permitted them to withdraw from it.

B. Arousal-Reducing Properties of Clinging

If clinging constitutes a form of psychological withdrawal from a distressing situation, it should be possible to show that an aversive stimulus is rendered less effective if presented to a subject while it is engaged in clinging. This hypothesis was tested in two experiments with three neonatal chimpanzees (Mason & Berkson, 1962). Both experiments measured distress vocalizations (whimpering, screaming) caused by shock to the foot. Shocks were administered when the infant was held in the ventroventral position and when it rested on a bare surface. The first experiment involved repeated shocks given at progressively higher intensities. Figure 16 shows that vocalizations remained at a uniformly low level at all magnitudes of shock when the infant was held, whereas crying increased steadily with increasing shock when the animal rested on a bare surface. Figure 17 shows a similar outcome, in this case using

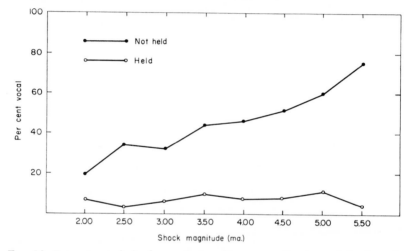

Fig. 16. Percentage of shocks producing distress vocalizations while the subjects were held and while they were resting on a bare surface. (After Mason & Berkson, 1962.)

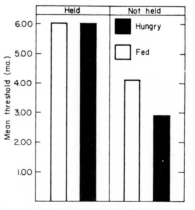

Fig. 17. Thresholds for vocalization to shock in relation to holding and food deprivation.

a measure of threshold based on the level of shock required to elicit vocalization. This experiment also showed that food deprivation lowered the threshold for vocalization only when the subject was not held.

Similar results have been obtained in experiments with 12 African-born animals. The subjects were tested individually in a strange environment, and the effects of seven conditions of social stimulation on their distress vocalizations were recorded. The seven conditions were: (H) holding in the ventroventral position described earlier, (G) grooming, (T) tickling (play), (C) clasping the subject's hands, (V) placing of palms of hands against the subject's ventral surface, (A) stimulus-person standing passive 2 feet away from the cage, and (O) an out-of-sight condition in which the subject was entirely alone. Conditions H, T, and G were of primary interest. Conditions C and V were introduced in order to determine whether different components of the clinging pattern might be differentially effective in reducing vocalizations. Conditions A and O served as controls. The primary dependent variable was the percentage of the 60-second test period in which vocalization occurred. The results shown in Fig. 18 indicate that condition H was maximally effective in reducing vocalization; G ranked second; C and V were of intermediate effectiveness; and T and A were only slightly more effective than O, no social stimulation. In a second experiment, the five conditions involving bodily contact (H, G, T, C, and V) were presented in a novel situation as in the preceding experiment, except that the subject was able to approach or withdraw from social contact. The results show a positive correlation between effectiveness of a social stimulus in reducing distress vocalizations and its attractiveness to the subject as measured by the percentage of 15-second intervals spent in proximity to the stimulus-person during stimulation (Fig. 19).

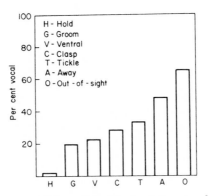

Fig. 18. Effects of various forms of social stimulation on distress vocalizations in an unfamiliar situation.

It will be recalled that, following a period of separation, cagemates displayed a characteristic sequence of activities in which clinging appeared first, followed by grooming, and finally by play. This sequence presumably reflects a gradual reduction in arousal level, but it is not clear from these data whether or in what manner the various patterns in the sequence bring about such a change. We have shown, however, that grooming a distressed chimpanzee or permitting it to cling reduces arousal as measured by vocal responsiveness. One might expect, therefore, that the social interaction that occurs between cagemates when they are rejoined following a period of separation could be modified by

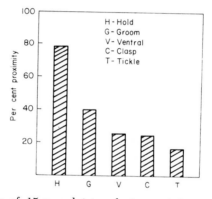

Fig. 19. Percentage of 15-second intervals in proximity to stimulus-person presenting various forms of social stimulation in an unfamiliar situation.

providing different forms of social stimulation before the animals are reunited. Such an experiment was conducted on eight pairs of African-born animals. Cagemates were separated for 90 minutes. Immediately before they were brought together in an unfamiliar room, each animal was played with or held or groomed by a stimulus-person for 1 minute. As a fourth condition, no social stimulation was provided (Alone). The animals were observed for 15 minutes after they were reunited. The sequence of response patterns was the same as that obtained in the original separation experiment, and it was found that the incidence of clinging was lowest following holding and increased progressively following grooming, play, and no stimulation (Fig. 20). The opposite trend was found for social play. The frequency of social grooming was low throughout, and there was no indication that it was systematically related to the experimental conditions.

Recent investigations with human infants suggest that sucking may have effects on arousal which are similar to those produced by clinging and being groomed (Bridger, 1962; Kessen & Mandler, 1961).

C. Arousal-Producing Properties of Play

The rough-and-tumble character of play, entailing bursts of violent motor activity and frequent changes in patterns of stimulation through various sensory modalities, suggests that it is an activity which heightens arousal. Some support for this expectation was found in the experiment described in Section II, B (Mason *et al.*, 1962). It will be recalled that in this experiment the stimulus-person resumed a passive role after interacting with the subject. Table I permits comparison of social responses during the initial passive period, during social stimulation, and during

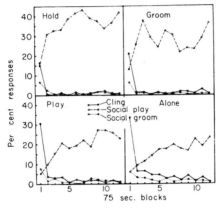

FIG. 20. Effects on social interaction between cagemates of providing different forms of stimulation before animals are reunited.

the passive period following stimulation. Percentage scores are presented for proximity (also see Fig. 6), and for "reaching" through wire, "grasping" stimulus-person, "pressing" trunk against wire, "panting" (rapid multiple exhalations of breath), "grooming" stimulus-person, making "lip movements" (smacking or twitching of lips), and "manipulating self" (scratching, picking). Inspection of Table I reveals that when the stimulus-person resumed a passive role after engaging in play with the subject, all measures of social responsiveness increased compared with either the initial passive period or with the terminal periods for the other social activities. This is consistent with the view that play increases arousal.

An increase in arousal when the level is already high should have aversive consequences (Hebb, 1958). Accordingly, one would expect that the incidence of spontaneous play, or of contact with a person providing play stimulation, would be low under conditions of high arousal. The evidence that has been presented indicates that this is the case. One might also expect that stimulation from play would combine with other sources contributing to arousal. In an effort to test this possibility we contrasted the arousal-producing properties of visual novelty and of play stimulation. Seven laboratory-reared chimpanzees were given paired-comparison tests with a stimulus-person in a familiar costume, which had always been associated with play, and with a stimulus-person in a novel costume. During the first twenty 150-second sessions, neither person responded to contact; in the next twenty, both responded with mild play; in the next twenty, with play of standard

TABLE I

PERCENTAGE OF 15-SECOND INTERVALS IN WHICH EACH RESPONSE OCCURRED
DURING INITIAL PASSIVE PERIODS (PASSIVE I), DURING SOCIAL STIMULATION
(ACTIVE), AND FOLLOWING SOCIAL STIMULATION (PASSIVE II)[a]

Phase	Response	Social Stimulus				
		Play	Pet	Groom	Present for Grooming	p^b
Passive I	Body press	4.8	4.0	3.0	7.0	—
	Grasp S-P	8.2	7.6	9.6	7.0	—
	Groom S-P	9.8	9.8	6.2	9.8	—
	Lip movements	9.8	12.0	6.4	11.0	—
	Manipulate self	33.2	40.2	39.0	42.6	—
	Pant	0.4	0.6	0.6	0.2	—
	Proximity	44.4	41.6	41.4	41.6	—
	Reach	22.8	16.4	18.2	19.2	—
Active	Body press	49.4[c]	33.6	28.8	9.8	.01
	Grasp S-P	62.6[c]	27.8	11.0	11.6	.05
	Groom S-P	0.6	8.8	3.8	42.4[c]	.01
	Lip movements	0.8	20.4	13.2	41.4[c]	.01
	Manipulate self	4.0	24.2	15.4	29.0[c]	.01
	Pant	86.6[c]	18.4	7.4	0.0	.01
	Proximity	95.8[c]	81.4	86.6	61.2	.05
	Reach	22.0	15.0	17.0	25.2[c]	—
Passive II	Body press	11.7[c]	4.4	4.0	7.2	—
	Grasp S-P	13.0[c]	1.2	4.2	5.0	.04
	Groom S-P	21.0[c]	6.0	4.5	4.8	.03
	Lip movements	19.4[c]	10.6	7.0	8.2	—
	Manipulate self	30.6	61.8[c]	47.0	46.0	.01
	Pant	1.2[c]	0.0	0.0	0.0	—
	Proximity	57.4[c]	29.8	32.9	27.8	—
	Reach	28.8[c]	8.6	11.2	9.8	.05

[a]From Mason *et al.* (1962).
[b]Friedman test.
[c]Activity in which response occurred most frequently.

vigor; and during the final twenty sessions, with vigorous play. A similar descending series terminated the experiment. The number of 15-second intervals in which contacts occurred decreased with increasing intensity of play, and this is consistent with the thesis that play enhances arousal. Of greater interest in the present context, however, is the finding that, as play intensity increased from mild to vigorous, the percentage of contacts with the novel stimulus-person decreased progressively (Fig. 21), suggesting that the choice of play stimulus-person

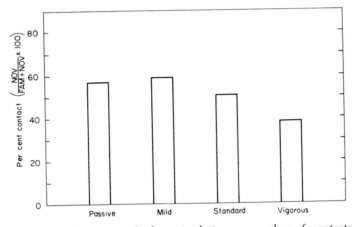

Fig. 21. Effects of intensity of play stimulation on number of contacts with a novel social stimulus (NOV) expressed as a percentage of total social contacts. FAM: Familiar Stimulus.

was jointly determined by visual novelty and intensity of play. Presumably, as the intensity of play approaches aversive levels there is a tendency to shift to the more familiar, hence less arousing stimulus.

We have evidence suggesting that the arousal-producing effects of play stimulation are present in early infancy. In an experiment with neonatal chimpanzees, we have found that signs of heightened arousal may be induced by patterns of mild cutaneous and kinesthetic stimulation which elicit vigorous play under appropriate conditions with somewhat older animals. The subjects were tested while supine on a bare table. The experiment measured their distress vocalizations in response to stimulation of the abdomen by tickling, rhythmic stroking, steady pressure, and a cloth-covered cylinder; in response to manipulation of the limbs; and in response to restriction of spontaneous movements of the head and of the limbs. The animals were tested three times a week from the first week to the eighth week of life. Table II shows the incidence and duration of vocalizations during stimulation. Manipulating the limbs, tickling, and restricting spontaneous movements produced a substantial increase in distress vocalizations as compared to the control condition (supine on table, no stimulation). Continuous light pressure on the abdomen and stroking produced no appreciable change in vocalizations compared to the control condition, and placing a cloth-covered cylinder on the subject's ventral surface caused a pronounced decrease in vocal activity. The cylinder closely approximated the postural adjustment during clinging.

TABLE II

MEAN DURATION OF VOCALIZATIONS PER SESSION
(IN SECONDS), AND PERCENTAGE OF SESSIONS IN
WHICH VOCALIZATION OCCURRED[a]

Treatment	Per cent sessions	Mean duration
Manipulate Limbs	80.2	8.8
Tickle	67.7	6.3
Restrain Head	62.5	8.8
Restrain Limbs	52.1	4.7
Pressure	43.7	3.8
Stroking	38.5	3.0
Control	43.7	2.6
Cylinder	14.6	0.5

[a]Data are based on twenty-four 30-second test sessions for each of four subjects.

V. CONCLUDING REMARKS

The data presented in this chapter show that regular variations in certain patterns of social responsiveness may be induced in young chimpanzees by a variety of procedures which seem to have in common the characteristic of altering the general level of excitement. I have ascribed the changes thus induced to variations in a hypothetical level of arousal, following the general conception developed by Bindra (1959), Duffy (1951), Hebb (1955), Lindsley (1951), Malmo (1959), and others. The arousal concept has at least three attributes which recommend its application to the present findings: 1. *Nonspecificity*. Our results indicate that many variables, including intense auditory stimulation, strange surroundings, separation from a cagemate, physical restraint, and novel social stimuli, may have equivalent effects on social responsiveness, suggesting a common mechanism which is not closely linked to specific stimulus attributes or modes of stimulation. Variations in level of arousal are generally assumed to be nonspecific in origin.

2. *Qualitative variations in behavior as a function of quantitative change.* The arousal concept offers the possibility of explaining changes in patterns of social activity without recourse to specific, independent drives, e.g., for play, clinging, and grooming. Our evidence suggests that an orderly transition from one social pattern to another will occur as the result of progressive changes in arousal level induced, for example, by habituation procedures or by an increase in the level of external stimulation. It should be noted, however, that the presumed function of the level of arousal is to establish a *predisposition* to engage in certain patterns of behavior, rather than to serve as a precipitating cause (Nissen, 1953, 1954). Behavior is presumably jointly determined by arousal level and by the presence of appropriate stimulus objects. Thus we would expect that an animal disposed toward social play would, in the absence of suitable companions, engage in strenuous motor activities such as swinging, jumping, and rolling, or would interact vigorously with objects in the physical environment. These activities will cease at levels of arousal which would induce clinging in a social situation, and the solitary animal may then engage in huddling and self-clasping (Fig. 2) or show intense vocal and escape activities (Menzel *et al.*, 1963). 3. *Reinforcement.* It has been suggested that changes in arousal level are reinforcing (Hebb, 1958), and attempts to alter instrumental behavior by direct stimulation of the arousal system in the brain have met with at least limited success (Glickman, 1960). Our data indicate that social activities may serve as rewards for the performance of simple instrumental responses, but they also show that the reinforcing effect of a given social activity depends on motivational state (see also McCulloch, 1939). Generally speaking, we have found that social play is sought in conditions of low or moderate arousal and is avoided when arousal is high, whereas the reverse is true of clinging. Moreover, we have evidence that play and clinging appear to augment and reduce arousal, respectively, suggesting that an important source of the reinforcing effects of these activities is their ability to maintain arousal within an optimum range (Leuba, 1955; Young, 1955).

The concept of arousal has recently been found useful in interpreting experiments on exploratory behavior and object manipulation (Welker, 1961), on sexual behavior (Beach, 1956), and on sucking activities of the human neonate (Bridger, 1962). Whether in fact the same mechanism is involved in each of these behaviors can be determined only by further research. Until such information is available, however, the arousal concept provides a way of dealing with many commonly-observed behavioral relationships that might otherwise be regarded as anomalies of little or no theoretical importance. This may be especially true for the primates. Among chimpanzees, for example, the excitement associated with the approach of mealtime may be expressed in social play, in fighting, or in increased sexual activity. Also, the frequency and intensity of social interaction between older chimpanzees is generally greatest just after the animals have been introduced and then diminishes progressively (Crawford, 1942; Nissen, 1951).

It is reasonable to assume that these effects are reflections of heightened arousal, but the task of determining whether such behaviors as copulation, fighting, and social play can be differentiated in relation to the level of arousal at which they characteristically occur will be a difficult one. Most likely some behaviors, e.g., sexual behavior and fighting, will show considerable overlap in their relation to arousal level. In this event, other factors such as specific cues, individual history, and the character of the social relationship may be critical in determining the form of the response.

REFERENCES

Beach, F. A. (1956). Characteristics of masculine "sex drive." *In* "Nebraska Symposium on Motivation" (M. R. Jones, ed.), pp. 1–32. Univer. Nebraska Press, Lincoln, Nebraska.

Bernstein, I. S. (1962). Response to nesting materials of wild born and captive born chimpanzees. *Anim. Behav.* **10**, 1.

Bindra, D. (1959). "Motivation: A Systematic Reinterpretation." Ronald, New York.

Bingham, H. C. (1927). Parental play of chimpanzees. *J. Mammal.* **8**, 77.

Bingham, H. C. (1928). Sex development in apes. *Comp. Psychol. Monogr.* **5**, No. 1 (Serial No. 23).

Bridger, W. H. (1962). Ethological concepts and human development. *Recent Advanc. biol. Psychiat.* **4**, 95.

Crawford, M. P. (1942). Dominance and the behavior of pairs of female chimpanzees when they meet after varying intervals of separation. *J. comp. Psychol.* **33**, 259.

Duffy, Elizabeth (1951). The concept of energy mobilization. *Psychol. Rev.* **58**, 30.

Falk, J. L. (1958). The grooming behavior of the chimpanzee as a reinforcer. *J. exp. Anal. Behav.* **1**, 83.

Glickman, S. (1960). Reinforcing properties of arousal. *J. comp. physiol. Psychol.* **53**, 68.

Harlow, H. F. (1960). Primary affectional patterns in primates. *Amer. J. Orthopsychiat.* **30**, 676.

Harlow, H. F., & Zimmermann, R. R. (1958). The development of affectional responses in infant monkeys. *Proc. Amer. phil. Soc.* **102**, 501.

Hayes, K. J., & Hayes, Catherine (1951). The intellectual development of a home-raised chimpanzee. *Proc. Amer. phil. Soc.* **95**, 105.

Hebb, D. O. (1955). Drives and the c.n.s. (conceptual nervous system). *Psychol. Rev.* **62**, 243.

Hebb, D. O. (1958). "A Textbook of Psychology." Saunders, Philadelphia, Pennsylvania.

Hebb, D. O., & Riesen, A. H. (1943). The genesis of irrational fears. *Bull. Canad. psychol. Ass.* **3**, 49.

Jacobsen, C. F., Jacobsen, Marion M., & Yoshioka, J. G. (1932). Development of an infant chimpanzee during her first year. *Comp. Psychol. Monogr.* **9**, No. 1 (Serial No. 41).

Kellogg, W. N., & Kellogg, Laverne A. (1933). "The Ape and the Child." McGraw-Hill, New York.

Kessen, W., & Mandler, G. (1961). Anxiety, pain, and the inhibition of distress. *Psychol. Rev.* **68**, 396.

Köhler, W. (1927). "The Mentality of Apes." Harcourt, Brace, New York. (Reprinted by Humanities Press, New York, 1951.)

Leuba, C. (1955). Toward some integration of learning theories: the concept of optimal stimulation. *Psychol. Rep.* **1**, 27.

Lindsley, D. B. (1951). Emotion. *In* "Handbook of Experimental Psychology" (S. S. Stevens, ed.), pp. 473–516. Wiley, New York.

McCulloch, T. L. (1939). The role of clasping activity in adaptive behavior of the infant chimpanzee. III. The mechanism of reinforcement. *J. Psychol.* **7**, 305.

Malmo, R. B. (1959). Activation: A neuropsychological dimension. *Psychol. Rev.* **66**, 367.

Mason, W. A., & Berkson, G. (1962). Conditions influencing vocal responsiveness of infant chimpanzees. *Science* **137**, 127.

Mason, W. A., Hollis, J. H., & Sharpe, L. G. (1962). Differential responses of chimpanzees to social stimulation. *J. comp. physiol. Psychol.* **55**, 1105.

Menzel, E. W., Jr. (1963). The effects of cumulative experience on responses to novel objects in young isolation-reared chimpanzees. *Behaviour* **21**, 1.

Menzel, E. W., Jr., Davenport, R. K., Jr., & Rogers, C. M. (1961). Some aspects of behavior toward novelty in young chimpanzees. *J. comp. physiol. Psychol.* **54**, 16.

Menzel, E. W., Jr., Davenport, R. K., Jr., & Rogers, C. M. (1963). Effects of environmental restriction upon the chimpanzee's responsiveness in novel situations. *J. comp. physiol. Psychol.* **56**, 329.

Nissen, H. W. (1931). A field study of the chimpanzee. *Comp. Psychol. Monogr.* **8**, No. 1 (Serial No. 36).

Nissen, H. W. (1951). Social behavior in primates. *In* "Comparative Psychology" (C. P. Stone, ed.), 3rd ed., pp. 423–457. Prentice-Hall, New York.

Nissen, H. W. (1953). Instinct as seen by a psychologist. *Psychol. Rev.*, **60**, 287.

Nissen, H. W. (1954). The nature of the drive as innate determinant of behavioral organization. *In* "Nebraska Symposium on Motivation" (M. R. Jones, ed.), pp. 281–321. Univer. Nebraska Press, Lincoln, Nebraska.

Nowlis, V. (1941). Companionship preference and dominance in the social interaction of young chimpanzees. *Comp. Psychol. Monogr.* **17**, No. 1 (Serial No. 85).

Schiller, P. H. (1952). Innate constituents of complex responses in primates. *Psychol. Rev.* **59**, 177.

Schiller, P. H. (1957). Manipulative patterns in the chimpanzee. *In* "Instinctive Behavior" (Claire H. Schiller, ed.), pp. 264–287. International Universities Press, New York.

Schneirla, T. C. (1959). An evolutionary and developmental theory of biphasic processes underlying approach and withdrawal. *In* "Nebraska Symposium on Motivation" (M. R. Jones, ed.), pp. 1–42. Univer. Nebraska Press, Lincoln, Nebraska.

Tomilin, M. I., & Yerkes, R. M. (1935). Chimpanzee twins: behavioral relations and development. *J. genet. Psychol.* **46**, 239.

Welker, W. I. (1956). Some determinants of play and exploration in chimpanzees. *J. comp. physiol. Psychol.* **49**, 84.

Welker, W. I. (1961). An analysis of exploratory and play behavior in animals. *In* "Functions of Varied Experience" (D. W. Fiske & S. R. Maddi, eds.), pp. 175–226. Dorsey, Homewood, Illinois.

Yerkes, R. M. (1933). Genetic aspects of grooming, a socially important primate behavior pattern. *J. soc. Psychol.* **4**, 3.

Yerkes, R. M. (1943). "Chimpanzees: A Laboratory Colony." Yale Univer. Press, New Haven, Connecticut.

Yerkes, R. M., & Tomilin, M. I. (1935). Mother-infant relations in chimpanzee. *J. comp. Psychol.* **20**, 321.

Young, P. T. (1955). The role of hedonic processes in motivation. *In* "Nebraska Symposium on Motivation" (M. R. Jones, ed.), pp. 193–238. Univer. Nebraska Press, Lincoln, Nebraska.

Chapter 10

Ontogeny of Perception[1]

Robert L. Fantz

Department of Psychology, Western Reserve University, Cleveland, Ohio

I. INTRODUCTION

A. The Problem

The behavior of adult primates is nicely adjusted to the environment so that it proceeds with few errors of spatial orientation, object recognition, or social perception. This might seem to indicate that characteristics of the environment are transmitted directly through the sense organs to the brain, and that the problem of perceptual development is simply to know when and under what conditions the transmission begins. But there is no way for the nervous system to directly "take in" any object quality or dimension; instead, a perceptual model of selected features of the environment must be created out of the patterns of incoming neural impulses, along with the intrinsic neural organization; this model must represent, in some coded fashion, enough of the physical character-

[1] The preparation of this chapter and some of the studies were supported by grant M-5284 from the National Institutes of Health, U. S. Public Health Service. Other previously unpublished work was supported by grant M-2497 from the National Institutes of Health and grant G-21346 from the National Science Foundation.

istics of the environment to mediate accurate and appropriate responses. The real problem, then, is to determine the origin of this schematic neural-perceptual model and of its close correspondence with behaviorally-important features of the environment.

The central importance of this problem is argued by Hayek (1952), who also points out the frequent neglect or denial of the problem by behavioral scientists. It is most easily neglected in the study of mature, experienced subjects whose perceptual model already corresponds well enough to the environment to mediate accurate and appropriate responses, and is sufficiently similar to the experimenter's own model to make these responses explicable and predictable. But often in the very young animal the perceptual model is not well developed, or cannot effectively be used to direct overt behavior; here the experimenter's perceptions are less helpful, and the need to determine the nature and origin of the subject's own view of the world is most apparent. And yet in the early stages of development the animal's perceptions are least accessible to experimental study.

In primates, or at least in apes and humans, the difficulty of testing perception at an early age is accentuated by the slow development of coordinated behavior involving the body and limbs, compared to most other animals. It is necessary to find a way around this difficulty, however. The long period of relative helplessness, and the long opportunity for perceptual development in primates before perceptual capacities can be used in adaptive behavior (Portmann, 1962), may be significant in several ways. It may be related to the eventual ability of many primates to react to more complex, subtle, and varied stimuli than do other animals. It provides increased opportunity for perceptual learning, which is thought to play a greater role than for nonprimates. And it brings the question of whether perception is basically learned or unlearned to a sharp focus in studying primates.

B. The Nativism-Empiricism Controversy

The nativist claims that the high degree of correspondence between the perceived environment and the physical environment is largely a result of the millions of years of development of the species under essentially similar environmental conditions, during which maladaptive sensory structures, neural connections, and response patterns have dropped out and those giving the closest correspondence have been retained. The empiricist argues that the correspondence arises largely during the postnatal development of each individual through cumulative experiences in a particular environment. These seemingly-opposite positions both seek the determinants of perception in the consequences of past behavior—either survival of the species or learning in the individual.

This emphasis on historical, environmental factors results in the frequent neglect of intraorganismic factors which are the immediate determinants (see Section III, B).

Visual perception, especially pattern and spatial vision, has been the main focus of the nativism-empiricism controversy, and it is the main area of study in this chapter. Research and arguments on both sides have been reviewed by Hebb (1949), Pratt (1950), Riesen (1958), and Zuckerman and Rock (1957). Some current approaches, which have given substantial but conflicting evidence, will be mentioned briefly.

1. EVIDENCE FOR EMPIRICISM

When students of animal learning turned to the area of perception, they made use of their well-developed training techniques. Classical conditioning became sensory-threshold determination, and instrumental conditioning became discrimination training, with little change except better control of stimuli. Eventually the discrimination-training method was applied to the study of perceptual development, largely through the inspiration of Hebb's (1949) experiments and theories. Since most newborn laboratory mammals are not trainable, animals were reared in darkness or other conditions of sensory deprivation until mature enough to be easily trained. It was hoped in this way to postpone perceptual development until testing was feasible and then to determine the effects of perceptual learning by comparing animals of the same age with and without visual experience.

The general findings of the visual-deprivation studies are well known (Beach & Jaynes, 1954). The initial deficiency in visual performance and the need for a period of visual experience appeared to give clear evidence for the essential role of learning, especially in the development of form and depth perception. The visual deficit was present to some degree in all species studied, and was marked in chimpanzees (Chow & Nissen, 1955; Riesen, 1958, 1961a).

Infant chimpanzees (*Pan*) reared for 7 months in complete darkness, or with 90 minutes daily of diffuse light, or of diffuse light with a moving fixation spot, all required several weeks or months to approach a nursing bottle (Riesen, 1958). One to 2 months were required to discriminate the bottle from other objects, or to show fear of a strange object. These responses were, of course, present in normally-reared animals. In discrimination training (Riesen, 1961a), presentation of a large, round plaque with vertical yellow stripes was followed by shock, while food was given after presenting any of four other plaques. Each of the four food plaques differed from the shock plaque in one respect: pattern, color, form, or size. Three animals, reared with only diffuse light until 7 months of age, made a total of 533 errors (approach to negative or withdrawal from positive plaques) when run to a criterion of 18 consecutive errorless trials, compared with 205 errors by three control animals reared with patterned stimulation. The deprived chimpanzees withdrew from the horizontally-striped and the red food plaques no more often than did the control animals, but they approached the shock plaque and withdrew from the small food plaque three times as often as the controls, and withdrew from the square food plaque five times as often. These and other data show that "the difficulty of discrimination of form is greatly reduced for the visually naive subject if the differences between the stimuli to be discriminated are replicated throughout the figure, as in horizontal vs. vertical striations, or in horizontal vs. vertical rectangles" (Riesen, 1961a, p. 69).

On the whole, then, visually-guided behavior of chimpanzees reared without patterned stimulation was inferior to that of normally-reared animals. These and other deprivation studies, especially those with kittens (Riesen, 1958, 1961a), have shown that patterned visual stimulation is important for normal visual development, and that, if it is not given early enough, it may be difficult for the animal to use vision effectively.

Empiricists usually attribute such visual deficits to lack of the long period of "primary visual learning" that they believe to be necessary for form perception (Hebb, 1949). However, other interpretations are possible (see Section III, A). For example, all of the visually-deprived chimpanzees showed oculomotor abnormalities (Riesen, 1958), and there was evidence of neural deterioration in some cases, which in one animal was eventually accompanied by blindness (Chow *et al.*, 1957).

2. EVIDENCE FOR NATIVISM

Direct evidence for nativism comes from studies by zoologists and ethologists of instinctive behavior in nonmammalian animals (Lorenz, 1957; Tinbergen, 1951). Some instinctive responses are released by certain sign stimuli, which are often configurational. Ethologists have claimed that the releasing mechanisms are genetically determined since, like other components of instinctive behavior, they are species-specific, stereotyped, and nonextinguishable. Others have questioned this, showing that in some cases the stimuli that release instinctive behavior sequences can be modified by experience, and that unlearned and learned factors may be woven closely together in the adult perceptions (Beach, 1960; Lehrman, 1953). Since species-specific learning experiences can cause species-specific responses, previous experience must be controlled to be certain that the discriminations are not learned. Studies of young animals and animals reared in isolation suggest a large repertoire of unlearned visual discriminations, including some discriminations of form differences (Thorpe, 1956; Tinbergen, 1951; Tinbergen & Kuenen, 1957). For example, herring-gull chicks, taken from the nest on the morning after hatching, pecked more at models of the parent's bill which were slender and elongated, which pointed down, and which had a protrusion near the end (Tinbergen & Perdeck, 1950).

Still better control over experience was achieved in experiments with domestic chicks. Soon after hatching, chicks can peck fairly accurately (Hess, 1956). According to observational reports, initial pecks are randomly directed toward all small, bright targets. But stimulus-preference experiments refute this. In the first minutes of visual experience, when presented a graded series of solid forms seen through plastic, chicks showed a steep gradient of pecking frequencies, favoring round over angular forms (Fantz, 1957). The roundness preference was also present among flat forms, even with large differences in size between round and angular outlines, and it was not disrupted by reversing the colors of figure and ground (Fantz, 1954). Round solid forms were also preferred to similar flat forms (Fantz, 1958a). Thus form and depth perception, generalization, and stimulus equivalence, all phenomena that pose problems for theories of behavior and neural organization in mature primates (Hebb, 1949), occur in birds without opportunity for learning.

Evidence favoring the nativistic position is not restricted to unlearned discriminations. Newly-hatched birds can learn to discriminate between visual stimuli with little previous experience (Fantz, 1954; Hess, 1962; Pastore, 1958, 1962), in comparison with the deficient performance of visually-deprived animals (Beach & Jaynes, 1954; Tucker, 1957). Even in neonatal primates, form discrimination and generalization have been proven by using training procedures adapted to the response capabilities of infant rhesus monkeys (see Chapter 11 by Zimmermann and Torrey).

3. POSSIBLE RESOLUTIONS

The apparent conflict between the data showing form discrimination in young birds without opportunities for visual learning and the data showing the lack of such discrimination in visually-deprived primates can be explained in several ways.

a. Phylogenetic difference. Unlearned perception could be present in birds and other animals whose behavior is largely instinctive, but not in primates and other animals with slower development and greater plasticity of behavior. This would mean that similar visual abilities of birds and primates originate in phylogenetic development in one case and in ontogenetic development in the other. But postulating different developmental processes in these different phylogenetic groups is opposed by Zimmermann's demonstration of visual discrimination learning by neonatal monkeys and by the research described in Section II, which obtained evidence on unlearned visual discriminations in young primates through methods similar to those used with newly-hatched birds.

b. Artifact of deprivation. The visual deficit shown by visually-deprived primates could be an artifact of the deprivation technique: a result of effects other than preventing visual learning (see Section III, A). It was hoped, in conducting the research described in Section II, that light would be thrown on this question by using the visual-deprivation technique as an adjunct to the study of normal development, and by comparing subjects reared in darkness not with normally-reared ones of the same age but with newborn or less-deprived ones. This approach was made possible by the development of testing techniques that could be used with newborn as well as older subjects and with deprived as well as nondeprived ones.

c. Methodological difference. Another explanation is that the discrepancy is not one of facts but of words and methods, that different definitions of perception or discrimination are implicit in different testing methods. Most of the data purporting to show that animals need a long period of experience to learn to see form are based on learned responses that indicate recognition of familiar objects, while the evidence for unlearned responsiveness to configurations is mainly from instinctive behavior. A confusion between "learning to see" and "learning to recognize objects" is also present in the interpretation of the initial deficit of people who have had congential cataracts removed to permit pattern vision for the first time (Hebb, 1949; von Senden, 1960). The deficit consists mainly in the difficulty of attaching names to particular objects and of learning to use vision to direct behavior. The many instances of noticing differences between forms and of describing configurational aspects of their visual fields have not been considered evidence for form perception merely because the patients do not give learned meanings to what they see (Wertheimer, 1951; Zuckerman & Rock, 1957). One such patient did attach the terms "round" and "square" to a ball and toy brick on first presentation, but usually much experience is required to name objects correctly on the basis of form.

This confusion results from the lack of clear distinction between the development of visual perception and the development of overt visually-guided behavior. The latter development usually requires learning in primates, in contrast to the more frequent instinctive behavior in other animals. When a learned response is used to test perception, it is hard to know whether or not the perception is also learned. To avoid the circularity of using learned indicators of perception to assess the role of learning in perception, the research described below used an unlearned indicator.

II. RESEARCH

A. The Differential-Fixation Method

The methods available for studying perception in newborn primates, especially apes and humans, are limited. There is no adaptive and discriminative response to objects comparable to the pecking of birds. The orienting response of the eyes and head provides the best solution since the eyes are open and active immediately after birth. Are the eye movements and fixations random, or are they selective and discriminative, favoring certain parts of the environment? They are certainly not completely random: the eyes of newborn chimpanzees and humans, and of chimpanzees reared in darkness, fixate a light source and follow a striped field. Eye movements completely independent of environmental stimulation under all conditions would be a sign of blindness. Empiricists generally admit initial reflex visual reactions and localization of a figure against the ground. The question is whether or not newborn primates can discriminate among stimulus targets, especially those differing in form, pattern, or depth.

The answer is given by the differential-fixation method, in which the response measure is the relative duration of visual fixation of different targets. Consistent differential fixation during exposures using the same targets in varied positions is termed a *visual preference*. Similar preferences shown with different sets of targets having a common stimulus characteristic may suggest a *visual interest* in that characteristic (e.g., a visual interest in pattern as a generalization from preferences for various patterned over plain targets). Visual preferences also provide one type of data on a more general process—*selective attention*—which may be important in perceptual development (see Section III, B).

B. Visual Preferences in Infant Chimpanzees

The differential-fixation method was first adapted for studying perceptual development in primates at the Yerkes Laboratories of Primate Biology. Three infant chimpanzees were taken from their mothers soon after birth, reared in a nursery with or without normal visual experience, and tested intensively for visual preferences. For the tests, the subject was restrained in a small hammock crib so as to keep its gaze on the ceiling of a test chamber. The inside of the chamber was uniformly white except for two stimulus targets on either side of a central observation hole. Through this hole either a camera or the experimenter recorded how long the chimpanzee fixated each of the targets during a series of 15-second test periods. Direction of gaze could be objectively determined, either directly or by frame-by-frame analysis of photographs, by observing the tiny images of the stimulus targets reflected from the cornea of the eye; fixation of one of the targets was indicated by the corresponding image being in the center of the cornea, over the pupil. Using this simple criterion it was possible to follow the gaze of the infant and record how long each target was fixated. Repeated independent analyses of photographic records showed that the response measure was reliable (Fantz, 1956).

The first subject, Pandit, was reared with unrestricted vision. In preliminary testing at 1 week of age he looked mostly at the contours of large figures, and he moved his eyes consistently in the direction of a line placed horizontally or vertically. At 1 month, when two smaller patterns were first presented together, a cross was fixated more than a circle in seventeen of twenty 15-second periods, while a square and a circle were fixated about equally. During the succeeding months, visual preferences were studied using a variety of stimulus pairs. In most cases the animal preferred the more complex of the two (Fantz, 1956).

The second subject, Mack, was visually deprived until 32 weeks old. He was given diffuse light for 90 minutes each day, but the rest of the time he was in complete darkness except for a flash of light, too short to permit change of fixation, every 8 seconds. When first tested after 1 hour of unrestricted vision, he showed strong spontaneous nystagmus which influenced the direction of gaze. Photographs of his eyes while exposed to a 22-inch-diameter black circle revealed the fixation point to be close to the top edge of the circle most of the time when his eyes were not moving (Fig. 1, left). When he was shown a black inverted triangle (equilateral with 25-inch sides) the fixation points clustered around the upper-right vertex. With the same triangle upright, fixations were more distributed around the figure but were still mainly at the top (Fig. 1, right). Gross eye movements were mostly parallel to the left side, with large saccadic movements down and irregular nystagmic movements up. So Mack's eye movements, though markedly abnormal, were related to the stimulus field.

When two smaller targets were paired either in right and left or in upper and lower positions, Mack fixated mostly the right or the upper position. Because of this oculomotor peculiarity, the targets were then presented in the preferred right and upper positions in a balanced sequence. With the technique thus adapted, the subject discriminated form and size. Out of eight test periods, a cross was preferred in six and a circle in one, with the cross fixated 67% of the time. Out of eight periods using 6-inch and 4-inch squares, the larger square was fixated more often in five and the smaller in one, with the larger square fixated 87% of the time. The following day, after 2 hours of unrestricted vision, Mack showed the same consistency of differential response to cross and circle but with the circle now preferred. Response to these forms was more variable in later tests.

After 1 week of restricted visual experience, Mack was tested with the normally-reared Pandit, then 24 weeks old. Table I shows the results for four pairs of targets, each presented four times on either three or five test days. Both animals gave evidence of visual discrimination, but the deprived infant showed more differential responding than did the normally-reared infant when the targets differed in size, and less when the targets differed in form (cross versus circle).

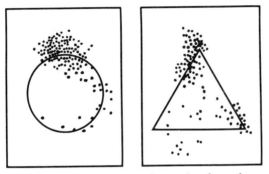

Fig. 1. Directions of gaze of a visually-deprived infant chimpanzee when exposed for 10 seconds to each of two black forms against a white ground. Points show positions of center of infant's pupil relative to reflection of target from cornea in successive frames of motion pictures.

If the reported visual deficit of visually-deprived chimpanzees is caused by incidental effects of visual deprivation, such as oculomotor abnormalities, rather than by lack of opportunity for visual learning, then a newborn infant might show more discriminative responses than a deprived subject. This was true of one newborn subject, Pandor (Fantz,

TABLE I

NUMBER OF TEST PERIODS IN WHICH A NORMALLY-REARED (N) AND
A VISUALLY-DEPRIVED (D) INFANT CHIMPANZEE SHOWED LONGER
FIXATION OF EACH MEMBER OF FOUR STIMULUS PAIRS

Subject	Cross-Circle	Large-Small	Red-Gray	Head-Oval
Pandit (N)	17–2[a]	13–7	8–4	8–3
Mack (D)	10–9	17–3[a]	8–4	9–1[a]
Combined	27–11[a]	30–10[a]	16–8	17–4[a]

NOTE: Cross-circle and large-small pairs were shown to each infant in 20 periods, red-gray and head-oval pairs in 12 periods. Periods with equal fixation were not included in data.

[a] $p < 0.05$ (two-tailed sign test).

1958b). From the day of his birth through the first 7 weeks, he showed consistent differential fixation in tests with various stimulus pairs. During this time he was in the dark except for 30 minutes or less in the test chamber and several minutes exposure outside the chamber daily. He usually preferred simple outlines or plain surfaces to more complex forms or patterned surfaces, in contrast to the results with the normally-reared subject. This suggests that *preference* for a highly patterned stimulus might require visual experience, even though the consistent favoring of the less-patterned target shows that the ability to *see* pattern was present from the first.

This possibility was given some support by the results of duplicate series of tests of the neonatal subject, Pandor, one series given just before the end of visual deprivation, starting at age 6 weeks, and the other after 2 weeks of unrestricted vision (Fantz, 1958b). For each of four pairs of targets (cross versus circle, sharp versus blurred photograph of human face, striped rectangle versus circle, checkerboard versus plain square), the one with the longer contours (first-named) was fixated more after the period of visual experience than before; in the first two cases this was a reliable reversal of preference from the shorter- to the longer-contoured form. The changes cannot be attributed to repeated testing experience since there was less change from day to day in each series of tests than after an interval of 2 weeks. The only reliable change in response during the period of deprivation from birth is illustrated in Fig. 2. Here also there is little indication of a specific learning effect of repeated testing since differential fixation first increased and then decreased under constant conditions of exposure. A more plausible interpretation is that the initial increase was due to improved ability to maintain fixation of the preferred target (circle), while the later decrease reflected increased tolerance for the complexity and contrast of the striped rectangle.

Other tests of these subjects are relevant to the question of possible effects of experience. Pandit, after 5 months of normal rearing with bottle feeding, failed to consistently differentially fixate the nursing bottle paired with either a cardboard cylinder or a large red-painted light bulb in tests given before feeding. Two months later, he and the two deprived subjects were given a series of tests, all before feeding, using a bottle versus the red-painted light bulb. Pandit looked somewhat more at the bottle (17 test periods versus 12 for bulb). Mack, after 3 months of visual experience with the nursing bottle, preferred the bulb (8 versus 22), as did Pandor after 1 month of experience (5 versus 15). The outcome for Pandit may have resulted from greater opportunity to associate the sight of the bottle with food; however, the effect was surprisingly small considering the length of time involved.

All three infant chimpanzees fixated a model of a human head more than a plywood oval. This preference was shown equally strongly by Mack 1 week after being moved from the darkroom and by Pandit after 6 months of unrestricted vision of people. Pandor, who lived and was fed in the dark, showed still greater responsiveness to the head model. Neither deprived subject showed increased responsiveness with experience. Apparently the head model was preferred not because heads had been associated with reinforcers, but because of intrinsic stimulus characteristics such as a complex and centralized pattern, color and brightness contrast, and solidity.

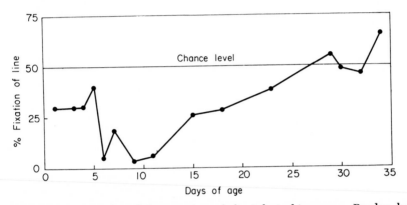

Fig. 2. Percentage of total fixation time of the infant chimpanzee, Pandor, kept in darkness when not tested, devoted to a 3-inch by 28-inch rectangle ("line") with 1-inch stripes of red, black, and blue, when paired with a 7-inch gray circle and presented in varied positions against a white ground. Each point is the mean for eight 15-second test periods.

These results suggest considerable innate perceptual capacity in primates, even after prolonged deprivation. They also reveal changes that may or may not indicate visual learning. In order to show the consistency of response among individuals, we must turn to the most numerous primate species.

C. Pattern Discrimination in Infant Humans

Neonatal humans, even less mature and coordinated than chimpanzees, are still able to direct their eyes toward selected parts of the environment. This indicator of visual discrimination was used in a series of studies. Pairs of patterns were presented for short periods against the ceiling of a test chamber. The infant's eyes were observed through a center peephole, and responses were recorded by the experimenters on timers. The direction of visual orientation was again indicated by the corneal reflection located over the pupil. Reliability was shown by similar data from two observers and in repeated tests (Fantz *et al.*, 1962).

In the first study (Fantz, 1958c), 22 infants were given ten weekly tests, starting at ages ranging from 1 to 15 weeks. There were three stimulus pairs varying either in form or in pattern, and a control pair of identical triangles. The control pair and a cross versus a circle did not elicit consistent differential responses. The other two pairs elicited highly reliable differential responses, with a checkerboard pattern fixated more than a plain square and concentric circles fixated more than horizontal stripes. An analysis by age (Fig. 3) indicates changes around 2 months of age. Before this time the square was fixated almost as much as the

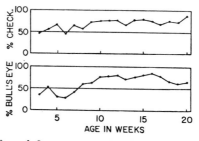

Fig. 3. Percentage of total fixation time of infant humans devoted to a checkerboard pattern paired with a plain square, and to concentric circles (bull's-eye) paired with horizontal stripes during two 30-second exposures. The positions of the targets during the second exposure were the reverse of those during the first exposure. Each of 22 infants was given ten weekly tests starting at varying ages; results for those tested at the same age were averaged. (From Fantz, 1958c.)

checkerboard, while the striped pattern was fixated reliably more than the concentric circles. These changes were not caused by repeated testing, since similar age differences were found on initial tests of these and 8 other subjects: 10 of 16 infants under 7 weeks preferred the checkerboard while 11 of 14 older infants did so; under 7 weeks only 2 of 16 preferred the circles over the stripes while all 14 older infants did so. Thus, the type of pattern was an important variable, as was patterning *per se*: differential response was greater between two patterns than between a patterned and a plain surface, especially during the first 2 months of age. The appearance of the checkerboard preference at 2 months was not caused by learning to see pattern, since the stripes and circles were discriminated earlier.

Further experiments were concerned with the pattern vision capacity of the infant human (Fantz *et al.*, 1962). Throughout the first 6 months of life, infants fixate a square of vertical black-and-white stripes more than a gray square of the same size and reflectance. We took advantage of this fact in order to test visual acuity. A series of squares with stripes of graded fineness was presented to determine the finest pattern that was fixated reliably longer than gray. Stripes subtending 40 minutes of visual angle could be discriminated by infants under 2 months, while stripes subtending 20 and 10 minutes were discriminated by age 4 and 6 months, respectively. More recent studies (Fantz, in press) have shown the resolution of still finer patterns under higher illumination by 2- to 6-month infants, and have shown the resolution of ⅛-inch stripes at a distance of 9 inches by infants less than 24 hours old. So even neonatal humans have considerable pattern vision, which indicates that their visual system is functional to some degree from cornea to cortex. The receptive and neural prerequisites for form perception are present in the primate soon after birth. Finer patterns are resolved at later ages, presumably because of further maturation of the eye and brain and increased skill in moving and focusing the eyes. These conclusions are also applicable to infant monkeys and apes, although their acuity at specific ages might be higher because they mature faster than humans (Jacobsen *et al.*, 1932; Ordy *et al.*, 1962).

The perception of solidity or depth was studied in other experiments (Fantz, in press). A white sphere and circle of 6-inch diameter were presented together with or without binocular depth cues, texture gradients, and shading, or with various combinations of these. A reliable preference for the sphere was shown only when the sphere had texture and brightness gradients, as do solid objects under most natural conditions. Binocular depth cues were not necessary for discriminating the solid object; in fact, under 3 months of age the differential response was greater with the left eye covered. Infants under 2 months did not consistently respond differentially. In another experiment, a solid but unpainted model of a human head was paired with a similar flat form. The solid object was strongly preferred by infants over 2 months old using either one or both eyes, but not by younger infants. Thus visual interest in the patterning accompanying solidity appears to require a period of maturation and/or visual experience. Whatever its origin, selective attention to solid objects provides differential experience with stimuli of future behavioral importance and gives the infant a chance to learn to discriminate depth cues other than patterning (see Section III, B).

The responsiveness of human infants to pattern and to various nonconfigurational stimulus variables was compared in other studies (Fantz, 1961, 1963). Six discs of 6-inch diameter were presented one at a time in repeated tests. The surfaces of three discs were patterned and those of the other three were colored uniformly. The length of the first fixation was measured in each case. Both newborn and older infants fixated the three patterned targets longer than the other three. Of the patterns, a schematic human face was preferred (Table II). None of the 43 infants looked longest at white, fluorescent yellow, or dark red. Thus, from birth on, length of fixation was determined less by color and brightness (often assumed to be primary and prepotent stimulus characteristics) than by pattern (previously thought to require either postnatal maturation or visual learning to be perceived). Similar patterns differing as much as five to one in area were fixated equally by infants over 1 month old, but under 1 month of age the larger patterns were preferred.

In summary, the results with infant humans support those with infant chimpanzees. They give stronger evidence for unlearned pattern discrimination and form perception in primates and show that direction of response is consistent among individuals. Infant humans in the early

TABLE II

MEAN PERCENTAGE OF TOTAL DURATIONS OF FIRST FIXATIONS IN
REPEATED PRESENTATIONS[a]

Age	N	Face	Circles	News-print	White	Yellow	Red	p[b]
Under 48 hours	8	29.5	23.5	13.1	12.3	11.5	10.1	0.005
2 to 5 days	10	29.5	24.3	17.5	9.9	12.1	6.7	0.001
2 to 6 months	25	34.3	18.4	19.9	8.9	8.2	10.1	0.001

[a]From Fantz (1963).
[b]Friedman analysis of variance by ranks.

months showed more differential responsiveness to patterns as opposed to plain or colored surfaces, outline form, and size than did the young or visually-deprived chimpanzees, although the comparison is only suggestive in view of differences in targets and other conditions. As with chimpanzees, infant humans revealed developmental changes which may or may not involve learning.

D. Effects of Length of Dark-Rearing on Infant Monkeys

A research program with infant rhesus monkeys (*Macaca mulatta*) was undertaken to give better-controlled data on the effects of visual experience and visual deprivation. This project, carried out at Western Reserve University, followed work at the Wisconsin Primate Laboratory that proved the usefulness of rhesus monkeys for studying the development of behavior (see Chapter 8 by Harlow and Harlow and Chapter 11 by Zimmermann and Torrey). The advantages in using monkeys are that their sensory capacities and behavior are similar to those of humans and apes, and that they are more mature at birth and continue to develop faster. So various types of visually-guided behavior, including learned responses, in addition to differential fixations, can be used to study the vision of rhesus monkeys, both immediately after birth and after rearing in darkness.

Most studies using visual deprivation have compared visually-deprived and visually-experienced animals of the same age. This confounds the prevention of visual learning and the degenerative effects of prolonged visual deprivation. To separate these factors, a number of infant monkeys were reared in darkness for different periods and then given the same tests, to bring out the effects of maturation under abnormal conditions without differences in opportunity for visual learning. These animals were compared further after rearing in a lighted nursery, to learn whether or not effects of visual experience were influenced by previous deprivation. It was hoped that the use of both learned and unlearned responses would help to clarify the role of learning in perceptual development.

1. Visual Preferences

a. Method. The testing procedure for infant chimpanzees and humans was modified to take into account the more advanced motor development, prone posture, and clinging tendency of the newborn rhesus monkey. When the monkeys were separated from their mothers, they were given, in their cages, terry-cloth-covered "mother surrogates" similar to those used by Harlow and Zimmermann (1959), except that the head was omitted for convenience in testing. The infant, clinging to the surrogate, was placed in a black cubicle (Fig. 4). The infant could look above the surrogate through a hole, 1 inch high by 4 inches wide, into a chamber 2 feet high, 2 feet wide, and 1.5 feet to the opposite wall. Two targets hung from this wall at the level of the infant's head and 1 foot apart. They were exposed by turning on the lights and dropping a shield covering the hole. The chamber was painted a nonglossy saturated blue as a contrasting background for the targets, which were illuminated by a 60-watt bulb on the chamber ceiling. Beams of light also came from either side of the infant's eyes to make the pupils more visible to the observer.

Sixteen pairs of stimulus targets were used. Many were taken from previous studies showing differential response by infant primates. Others were added to increase the range of stimulus variables. Differences among the pairs in color and other aspects were introduced to prevent decreased responsiveness during a test session. The results from the first four monkeys tested (Lady, Othello, Desy, and Portia) suggested that 13 of the stimulus pairs could be grouped into the four categories shown in Fig. 5, while the remaining three pairs (not included here) did not fit these categories and gave inconsistent results. This classification of stimulus characteristics gained support from the results for the other monkeys, which showed preferences and developmental changes that were often similar for the several pairs in a category. Thus, the designation of the critical stimulus difference was based on equivalences in the animals' responses as well as on the physical properties of the targets. It was hoped that this would help to bring out the animals' own basis for classifying stimuli (see Section I, A).

The 16 pairs were presented in random order usually twice a week, divided into three test sessions. The concentric circles versus stripes and "eye" spots versus off-center spots were used in two of the sessions to give additional data for these pairs. The test of each pair included two successive 15-second exposures with randomly-determined right and left positions for the first and reversed positions for the second; fixation times

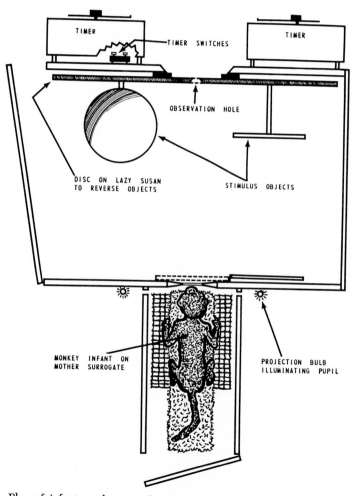

Fig. 4. Plan of infant-monkey visual-preference apparatus, with one of the stimulus pairs (6-inch sphere and circle) in testing position.

for the two were combined. As before, corneal reflections were observed through a center peephole and their durations were recorded on Microtimers.

The infants were placed in a lightproof room with their natural mothers within several days of birth. They were taken from the mothers at or before 10 days of age but continued to live in the darkroom for various lengths of deprivation. Testing started toward the end of deprivation in most cases and continued after deprivation until the animal could no longer be tested in the apparatus, usually around 5 months of age. Testing required about 5 minutes daily, with visual experience restricted to

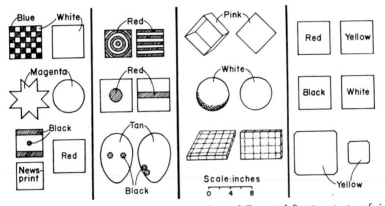

FIG. 5. Drawings of pairs of targets used in differential-fixation tests of infant rhesus monkeys. The four sections represent four categories of stimulus variation (left to right): (1) patterned versus plain: checkerboard—square, star—circle, black-and-white patterned square—larger red square, square of newsprint—larger red square; (2) centered versus uncentered patterns: concentric circles—horizontal stripes, circle—stripe, "eye" spots—spots at edge of ov̇ ⁻⁾̀ soᵢidity: box—square, sphere—circle, slanted—upright patterned surfaces; (4) nonconiigurational variables: red—yellow, black—white, 8-inch square—4-inch squɑ. ⁻ Each pair was presented against a blue background (the same as checkerboard color), with 12 inches between targets.

the inside of the test chamber. Occasional visual-cliff and other tests were also given to some animals during the deprivation period. If this minimal visual experience had any effect, it presumably would be to decrease the severity of the deprivation and the resulting differences among animals with different lengths of deprivation.

b. *Results after deprivation.* Table III shows the consistency of preferences in repeated tests for each of ten infants. This includes only tests given during the first 10 weeks of unrestricted experience. In each of the four patterned-versus-plain pairs, the infants deprived for less than 8 weeks all fixated the patterned target reliably more than the plain

TABLE III

NUMBER OF EXPOSURES IN WHICH FIRST OR SECOND TARGET OF EACH PAIR WAS FIXATED LONGER DURING FIRST TEN WEEKS OF UNRESTRICTED VISION[a]

Stimulus pair	Name of monkey and weeks of deprivation									
	Coco 0[b]	Osda 0.5	Tina 2	Lady 3	Benon 5	Waldo 5	Othello 6	Desy 8	Portia 11	Jean 16
Patterned										
Checks-square	**16–3**	**18–2**	**20–1**	**14–1**	**18–0**	**15–5**	**17–1**	9–7	**15–0**	11–8
Star-circle	**16–5**	**19–1**	**21–0**	**14–0**	**18–1**	**18–2**	**16–1**	11–2	**4–11**	**14–3**
Pattern-red	**19–2**	**18–2**	**17–4**	**14–2**	**18–1**	**16–4**	**13–5**	**2–12**	8–9	12–6
Print-red	**19–2**	**18–3**	**21–0**	**14–0**	**15–3**	**18–2**	**13–6**	**13–1**	7–12	10–8
Centered										
Circles-stripes[c]	27–22	**35–15**	**32–8**	**25–6**	19–16	20–20	21–16	17–13	16–16	16–20
Circle-stripe	14–7	13–7	**17–4**	**12–4**	11–8	12–7	12–7	7–10	**5–12**	**15–4**
Eyes-at edge[c]	10–18	**20–10**	**11–29**	**19–7**	21–16	15–25	18–17	**9–18**	**21–12**	15–19
Depth										
Box-square	14–4	**17–4**	**19–2**	**13–1**	**17–2**	**15–5**	**16–1**	**11–3**	**12–5**	**13–5**
Sphere-circle	14–6	14–6	**16–5**	10–4	13–6	12–8	7–11	7–7	5–10	12–5
Slanted-upright	12–7	12–7	15–6	11–5	9–10	**15–5**	**15–3**	5–10	12–6	10–8
Other										
Red-yellow	9–12	7–13	8–12	5–12	11–8	11–9	10–9	10–4	**13–2**	**14–4**
Black-white	**15–5**	**16–5**	**15–4**	9–4	**14–4**	13–5	11–6	**14–1**	**14–3**	**14–5**
8"–4" square	9–12	13–6	12–8	10–7	13–6	**14–5**	9–8	**16–1**	**15–4**	**15–3**

[a]Total for cells varies because of variations in number of exposures given and omission of tie scores. Boldface type indicates $p \leq 0.05$ according to two-tailed Wilcoxon T test of ranked differences between time scores for the two members of a pair.

[b]Coco left with mother in light and not tested during first month; data are from subsequent 10 weeks.

[c]Pair exposed more often than others.

target; after longer deprivation the preference decreased considerably and was reversed in several cases. Differential fixation of centered and uncentered patterns was less consistent, but was greatest for three of the less-deprived infants. All but the three longest-deprived infants preferred the three-dimensional objects fairly consistently. In contrast, these three showed more consistent color, brightness, and size preferences than the younger and less-deprived infants. So the effect of prolonged visual deprivation was not simply to reduce visual preferences in general; the effect depended on the nature of the stimulus variable, especially on the presence of configurational differences.

Another analysis shows the course of development during the early period of unrestricted experience (Fig. 6). Each graph shows changes with experience in the *mean* strength of preference for the generally-preferred targets (left member of each pair in Fig. 5) of the category; for the nonconfigurational category, the curves represent only the tendency to respond differentially in some consistent direction. Four-week periods of testing were averaged except for the first week of unrestricted experience, shown separately. The curves begin with the last 4 weeks (or less for some animals) of nearly-complete visual deprivation and

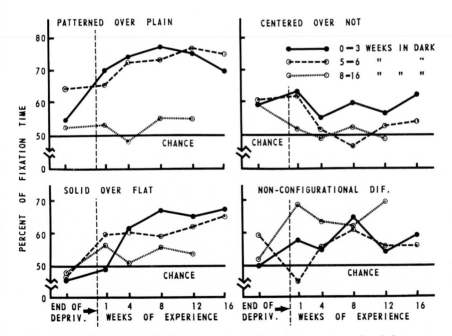

FIG. 6. Postdeprivation development of visual preferences in each of four categories of stimulus variables by groups of infant rhesus monkeys with different lengths of dark-rearing.

continue beyond the 10 weeks of experience included in Table III. The subjects were grouped in threes according to length of deprivation, omitting the one (Coco) not tested until after 1 month of experience. The subjects deprived for 8 weeks or longer showed little differential response to configurational variables at any time, while they responded differentially to nonconfigurational variables throughout the period of unrestricted experience, as was indicated in Table III. The two less-deprived groups fixated patterns two or three times as much as plain surfaces throughout the unrestricted experience; the increase during this time was not reliable. On the other hand, differential fixation of solid or spatially-projecting objects, as opposed to flat surfaces, increased reliably (analysis of variance by ranks) as a function of time spent in a normal lighted environment. Differential fixation of centered patterns and of nonconfigurational targets was variable and not related in any obvious way to the amount of visual experience.

These results suggest that visual experience increases visual interest in patterned stimulation, especially the pattern accompanying solid objects, but that the experience must come before 2 months of age to be effective. The greater differential response to nonconfigurational variables by animals deprived until this age indicates that the deprivation did not destroy their ability to discriminate visual stimuli in general.

c. *Results during deprivation.* The above results do not indicate the degree of responsiveness to visual configurations in the absence of *either* prolonged deprivation or unrestricted experience. Three of the above subjects (Benon, Waldo, and Jean) and three other monkeys were placed in darkness on the day of birth and tested once or twice a week during the first 5 weeks of deprivation. The amount of visual exposure as a result of these and other tests was about 1 or 2 hours a week per animal.

Figure 7 shows the mean degree of preference of these six infant rhesus monkeys as a function of age in weeks. Near-chance fixation of the three solid-versus-flat pairs of targets suggests that visual experience is crucial for developing a visual interest in solid objects, in agreement with results presented in Fig. 6. The three centered patterns were initially somewhat preferred to the uncentered patterns, and this preference was maintained without visual experience. The results for the four patterned-versus-plain pairs are graphed separately. The four curves, although at different percentage levels, are similar in showing a peak in the third or fourth week of age. Thus the visual interest in patterns increased during the first month whether or not the particular pattern was initially preferred to an unpatterned or colored surface. Comparison of Figs. 6 and 7 shows that this early upward trend with minimal visual experience continued if unrestricted visual experience was given before 2 months of age, but that with longer deprivation the downward trend starting at 1 month continued and was not reversed by later visual experience.

Further evidence for some unlearned visual preferences before the changes produced by maturation, prolonged deprivation, or visual experience, comes from analyzing all tests given during the first 2 weeks of age to those monkeys reared in darkness during this period (i.e., the six monkeys included in the results of Fig. 7 and three others with

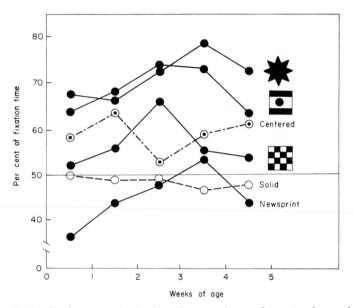

FIG. 7. Early development of visual preferences for configurational stimuli in infant rhesus monkeys kept in darkness except for testing. Broken lines are means for centered versus uncentered patterns and solid versus flat objects. Each solid line is for a patterned target paired with a plain one (see Fig. 5).

shorter deprivations). Of the 13 stimulus pairs, those showing a consistent direction of differential fixation among separate tests were concentric circles over stripes, "eye" spots over off-center spots, star over circle, black-and-white pattern over red, and yellow over red.

2. VISUALLY-GUIDED BEHAVIORS

A battery of tests was given daily to most of the rhesus monkeys used in the research described above. This testing program was usually begun the day before the subject was moved into the lighted nursery to live. Its purpose was to determine if length of deprivation would affect the use of vision in adaptive and learned behavior, and if the changes in visual preferences could be related to this behavior. Each test was continued until it was judged to be passed on the basis of repeated observations of performance, to provide a rough basis for comparing the subjects. This procedure and several of the specific tests were adapted from those used by Riesen (1951, 1958) to test chimpanzees reared in darkness.

a. Visual cliff. The tendency to avoid a spatial dropoff underneath glass has been used by Walk and Gibson (1961) to study the development of depth perception in many species of animals, including primates. The texture or patterning of the deep and shallow surfaces underneath the glass was found to be essential for the discrimination. This type of depth discrimination was shown on the first day of life by precocial birds and by ungulates, on first visual exposure by rats reared in darkness, after several days of visual experience by visually-deprived cats, and by infant humans as soon as they could crawl. Walk and Gibson concluded that these results support the hypothesis of innate depth perception. Two visually-experienced infant monkeys, 10 and 12 days of age, walked more on the shallow side of the cliff and locomoted abnormally when on the deep side, but did not completely avoid the deep side until a week or three weeks later; this was interpreted as indicating maturation of depth discrimination in this species. However, Rosenblum and Cross (1963), using a criterion of preference for the near side rather than avoidance of the deep side, found that visually-experienced infant rhesus monkeys discriminated depth at 3 days of age and thereafter.

The visual cliff used in the present study was a small version (3 feet square; 2.5 feet high) of Model III designed by Walk and Gibson (1961), with modifications. Panels hinged to either side of a board directly under the center of the glass provided a simple arrangement for reversing shallow and deep sides between trials and thus controlled for the position preferences sometimes present in monkeys reared in darkness. The experimenter was hidden from the subject by boards extending above the glass at front and back. Lights were placed 1.5 feet beyond transparent plastic sides of the apparatus, and were just high enough above the glass to give equal illumination of shallow and deep sides, but low enough to be reflected in the glass well away from the edge of the cliff. The hinged panels and the floor under and around the apparatus were covered with ¼-inch checkered plastic.

A subject was given six or eight trials a day when possible, with the position of the shallow side varied systematically. The procedure had to be varied among the subjects or even for a given subject because of differences in willingness to walk on the glass and in degree of disturbance in the situation. The starting place for the subject was sometimes a board or platform at one end of the cliff edge, with a mother surrogate at the other; it was sometimes a surrogate in the middle of the glass between shallow and deep; and it was sometimes a surrogate placed alternately on shallow and deep sides. In all cases the datum recorded was whether the monkey walked on the glass on the shallow or deep side, and the rough criterion of discrimination was a predominant choice of the shallow side during two or more consecutive test sessions. A more quantitative criterion was not feasible because of variations in procedure and in the behavior of the subjects; when the consistency of the response seemed at all doubtful, another session was given. In some test sessions the subjects did not walk on the glass under any condition so that no information was obtained until a later age.

In general, the longer the visual deprivation, the more visual experience required to prefer stepping onto the nearer patterned surface (Table IV). Infants deprived for less than 1 month required 4 days experience or less. Four subjects, including the three longest-deprived, did not avoid the cliff for as long as they could be tested; they continued to explore and play on both sides indiscriminately.

The differential response to near and far patterned surfaces underneath glass was clearly disrupted by prolonged visual deprivation. After deprivation for less than 1 month, perhaps the discrimination was innate or perhaps a short period of experience was required; this is unclear because the younger infants were initially uncoordinated and their responses ambiguous or variable so that several days were usually required to obtain sufficient data. Even Jubo, who reached criterion when first tested on the last day of deprivation, had about 3 hours of previous testing experience plus an unknown period of visual exposure on the morning of birth. Likewise, the newborn monkeys used by Rosenblum and Cross (1963) had 3 days of visual exposure before testing, while the the older subjects of Walk and Gibson (1961) had still more visual exposure. Thus the possibility is left open that monkeys require some opportunity for visual-motor experience to discriminate depth on the visual cliff, as was shown to be true for cats by Held and Hein (1963). They found that kittens reared in darkness avoided a visual cliff after receiving visual feedback from self-produced movements on a flat surface, and that similar visual stimulation received passively was ineffective. Assuming a similar experiential requirement in monkeys, the effect of prolonged visual deprivation in the present results may have been to interfere with receiving or utilizing the reafferent stimulation normally accompanying locomotion (see Section III, A).

b. Obstacle field. Spatial orientation was tested in a 40-inch-square box with patterned floor and sides and with 2-inch-square posts sticking up at 9-inch intervals in half the box. The infant was placed on its mother surrogate in the empty portion of the field and allowed to explore. It

TABLE IV

Number of Days of Normal Experience Required to Pass Visual Tests after Nearly Complete Visual Deprivation

| Test | Name of monkey and age at end of deprivation period | | | | | | | | | | |
	Osda 3 days	Jubo 9 days	Tina 2 weeks	Lady 3 weeks	Benon 5 weeks	Waldo 5 weeks	Othello 6 weeks	Desy 8 weeks	Portia 11 weeks	Jean 16 weeks	Bella 20 weeks
Visual cliff	4[a]	0[b]	3	2	31[c]	120[d]	19[c]	27	61[d]	72[d]	47[d]
Obstacle field	4	1	0[b]	1	1	0[b]	—[e]	3	1	21	30
Bottle location	3	1	2	4[c]	2	3	7	4[c]	1	7	14
Surrogate location	4	1	1	4[c]	2	1	—[e]	3	5	21	30
Bottle recognition	11	26	19	13	28	19	17	15	36	72	—[e]
Surrogate recognition	12	2	12	5[c]	14	11	—[e]	11	8[c]	58	34[d]
Mean, last 4 tests	7.5	7.5	8.5	6.5	11.5	8.5	(12.0)	8.2	12.5	39.5	(21.2)

[a] Chose shallow on 9 of 14 trials when first exposed to light; then disturbed and could not be tested until 4 days later.

[b] Reached criterion on last day of deprivation, after visual experience only during testing and with the mother immediately after birth.

[c] Reached criterion on first day test was given, or first day adequate test was possible.

[d] Testing discontinued even though criterion not met.

[e] Could not be tested because of disturbance, immobility, or poor motivation for task.

demonstrated spatial orientation by repeatedly walking between and around posts without touching them, by following walls and corners, and by reaching for posts and walls.

The two longest-deprived animals required 3 and 4 weeks of unrestricted visual experience to consistently avoid obstacles; all the others showed good spatial orientation after several hours or days of experience (Table IV). In no case was there evidence of immediate spatial orientation without any opportunity for learning, although this was hard to determine with the less coordinated and less mobile younger subjects.

c. *Object localization and recognition.* Tests using two tactually-familiar objects were begun, when possible, on the first day of unrestricted experience. An infant indicated localization of the 1- or 2-ounce nursing bottle by repeatedly reaching for the bottle with hand or mouth wherever it might be held. Then recognition was tested by presenting another small object beside the bottle and slightly closer to the infant. A decoy bottle was used on some trials to exclude olfactory cues. Correct choice with each of five objects was required for at least two test sessions. Comparable tests were given with the mother surrogate as goal object: first, to determine if the infant would walk directly to the surrogate placed in various positions in an open field; and second, to determine if it could discriminate the surrogate from five objects varying widely in form and color. Two of these objects at a time, and the surrogate, were arranged in varied positions on a flat surface, 2 feet away from the monkey's starting position and about 1 foot apart. Predominantly correct choices were required in two or more 6-trial sessions.

None of the four localization and recognition tests taken separately was consistently related to length of deprivation (Table IV), except that the infants deprived for 16 and 20 weeks needed more experience. However, taking all four tests together, the mean number of days to pass increased with length of deprivation, suggesting that visual deprivation retarded the development of object perception.

Not included in Table IV are the ages of appearance of two other observed responses—eyeblink to the edge of a rapidly-approaching hand, and cessation of eyeblink to touch of the monkey's cheek by a visible stimulus. These responses were sometimes ambiguous or variable from day to day, and the results showed no apparent relation to length of deprivation.

3. DISCRIMINATION TRAINING

a. *Method.* To determine any lasting effects of early visual deprivation, 14 of the monkeys used in the above tests were given discrimination training in a modified Wisconsin General Test Apparatus (see Chapter 1 by Meyer *et al.* in Volume I) at a mean age of 15 months. After training on a multidimensional problem, which all animals solved rapidly, they were given 24 unidimensional problems, each for 40 trials in a single session. There were four problems in each of six categories of stimulus variation. Three categories were taken from the visual-preference tests: (1) surface patterning or complexity of form, (2) centered versus uncentered patterns, and (3) solidity or orientation. The other three constituted the previous "nonconfigurational" category: (4) color, (5) brightness, and (6) size. Each category included stimulus pairs used in the visual-preference tests. The criterion for a learned discrimination was eight consecutive correct responses during the last 30 trials of the session ($p < 0.05$; Grant, 1947). The animals were compared both on the number of problems solved according to this criterion, and on the number of trials correct out of the last 30. The six animals deprived for 8 weeks or longer (the period after which visual preferences had changed markedly) were compared with the eight animals reared in darkness for shorter periods or not at all.

b. Results. Each of the less-deprived monkeys made a greater mean number of correct responses than any of the longer-deprived monkeys. The less-deprived group solved almost twice as many of the individual problems. The two groups of animals differed reliably ($p \leq 0.02$, Mann-Whitney U test) in number of trials correct for each of the six stimulus categories except color. Three of the long-deprived monkeys, with as much visual experience before testing as the less-deprived group, performed just as poorly as the other three, which had had a shorter period of unrestricted vision.

Comparison of results among the stimulus categories suggested that form and pattern were more easily discriminated than color, brightness, and size, especially by less-deprived monkeys. Criterion was met on a mean of 5 of the 12 problems in the first three categories, but the mean was only 2 of 12 in the three nonconfigurational categories. Upon closer analysis, however, the original classification of the discrimination problems was found to obscure a critical stimulus difference. The mean number of correct responses in the last 30 trials was greater than 20 in only 7 of the 24 problems; all 7 differed tactually (i.e., in form, texture, solidity, or size) as well as visually. Five problems were solved by 10 or more subjects, and, similarly, each involved a tactual stimulus difference. On the other hand, of the 14 monkeys, only 1, 0, and 3 met criterion with the seemingly gross but nontactual differences of checkerboard versus plain square, red versus yellow square, and white versus black square.

So the stimulus pairs were reclassified into four categories that cut across the original ones: (1) form—including star versus circle, sphere versus circle, box versus square, slanted versus level (all similar to those in Fig. 5), finely-textured versus smooth blocks, and broken-branch segment versus dowel-rod segment (similar in volume and both painted brown); (2) size—including a difference in thickness originally classified as solidity; (3) pattern—of painted plaques; (4) color and brightness. Table V summarizes the data. The difference between groups in number of trials correct was reliable for each category, although the superiority of the less-deprived group was greatest for the first two categories, which include tactual differences. The animals of both groups made more correct responses on problems involving form differences than on problems in each of the other three categories ($p \leqq 0.001$, Wilcoxon T test); size differences were also reliably easier than pattern or color-brightness, while the last two categories did not differ.

TABLE V

DISCRIMINATION PERFORMANCE OF YOUNG MONKEYS AFTER DARK-REARING
TO LESS THAN (L) OR MORE THAN (M) 7 WEEKS OF AGE

Category of stimulus difference[a]	Number problems given	Mean number problems solved		Mean number trials correct of last 30	
		Group L	Group M	Group L	Group M
Form	6	5.2	3.5	26.2	21.8
Size	5	1.6	0.6	23.1	19.0
Pattern	5	0.8	0.5	17.2	15.1
Color-Brightness	7	1.0	0.2	17.0	15.7
Combined[b]	24	9.0	5.0	19.9	17.0

[a]Revised classification of stimulus variables (see text).
[b]Combined data include a 24th object pair left out of categories because of an uncontrolled small tactual difference as well as the intended brightness difference.

It is clear that differences in visual form which can also be distinguished through touch or manipulation were most quickly discriminated by visually-experienced monkeys, and that the presence of a configurational difference in itself was not enough to make discrimination easy. One possible explanation is that learning was aided at first by touching the objects even though later correct choices were made without tactile comparison (i.e., the testing procedure permitted touching one object without pushing it aside and then shifting to the other object, but this rarely occurred in later trials of a problem). But at least two facts make this interpretation implausible. First, the long-deprived animals, whose deficit is presumably visual rather than tactile, were inferior on problems with tactual differences; second, touching does not help squirrel monkeys (*Saimiri sciureus*) to learn stimulus-object discriminations (Peterson & Rumbaugh, 1963). An alternative explanation will be presented in Section III, B.

c. Free-choice data. To see if there were stimulus preferences before training, all 24 stimulus pairs were presented randomly in varied positions for a total of ten times per pair. Response to either object was rewarded. This followed preliminary training to push aside a single block and familiarization trials with each of the object pairs in sight but out of reach. Using a criterion of more than twice as many choices of one object as the other, an initial preference was shown more frequently within pairs differing in form (61% of problems for all animals) than in size (29%), pattern (37%), or color-brightness (37%). This selective response to differences in form was more pronounced in the less-deprived group (69%) than in the long-deprived group (50%); for the other stimulus categories combined the groups did not differ as much (37% versus 30%).

For discrimination training, the positive object of each pair was designated independently of preference. Problems in which the positive object had been preferred were solved to criterion 48% of the time; problems in which the negative object had been preferred were solved 34% of the time; but when no consistent preference was shown before training, the problem was solved only 24% of the time. Thus stimulus preferences, as well as stimulus category and rearing conditions, are related to discrimination performance. With all three factors favorable—objects differing in form, subjects deprived for less than 8 weeks, and an initial preference for one object or the other—the objects were discriminated to criterion during a 40-trial session in 91% of problems.

E. Summary of Results

1. Visual Preferences

Differential visual fixations by infant chimpanzees (Section II, B), humans (Section II, C), and rhesus monkeys (Section II, D, 1) have proven that primates (a) have the visual structures and functions necessary for resolving fine patterns from birth, (b) can discriminate patterns without opportunity for visual learning, and (c) show strong unlearned visual preferences for particular stimulus targets starting at birth. Early preferences are generally related to configurational variables and tend to favor complex patterns, even over bright colors. Visual preferences persist throughout the early months either with or without visual experience, but show developmental changes: (a) visual interest in complex patterns or outlines increases during the early weeks with very little visual exposure; (b) this interest increases further, or is at least maintained, with unrestricted visual experience during the early months of life; (c) visual interest in solid objects appears after visual experience in the early months; (d) visual interest in complex patterns or outlines decreases sharply during prolonged visual deprivation; (e) unrestricted visual experience following prolonged deprivation does not bring about a consistent interest in patterns or solid objects; (f) preferences among targets differing in color, brightness, or size are most marked following prolonged periods of deprivation.

2. VISUALLY-GUIDED BEHAVIORS

Avoidance of a visual cliff, spatial orientation in an obstacle field, localization of the nursing bottle by hand or mouth, and approach to the mother surrogate were achieved within a matter of days by infant rhesus monkeys from 3 days to 5 weeks old when moved to a lighted nursery (Section II, D, 2). Discrimination of the bottle and surrogate from other objects required more visual experience. Little or no improvement was apparent from maturation alone. To the contrary, the older and longer-deprived animals tended to require longer in the light to show these various responses, or else did not respond for as long as testing was continued. This deterioration in visually-guided performance attributable to abnormal developmental conditions was evident earliest and most persistently in the visual-cliff test.

3. DISCRIMINATION TRAINING

After months of unrestricted visual experience, monkeys that had been visually deprived for less than 8 weeks learned discriminations consistently better than longer-deprived monkeys (Section II, D, 3). The superiority was most marked for stimulus pairs differing tactually as well as visually in form or size; these stimulus differences, especially form, were easier for all subjects to discriminate than were differences in pattern, color, or brightness. Before discrimination training, trials with either response rewarded had indicated frequent preferences within pairs differing tactually in form, especially in the less-deprived group. This suggested that a pretraining preference facilitated discrimination, even if the nonpreferred object was positive.

III. DISCUSSION AND THEORY

A. Possible Effects of Visual Deprivation

1. DELAYING VISUAL LEARNING

The dark-rearing technique was originally adopted as a way to delay visual learning until perception could be tested through training methods. The deficiencies of visually-deprived animals were then taken as showing the essential role of learning in perceptual development (Section I, B, 1). This interpretation has been questioned on various grounds.

The present demonstration that monkeys reared in darkness for 2 months or more are inferior to less-deprived monkeys both in unlearned visual responsiveness and in later learning (Section II, D) is direct evidence that visual deprivation has effects other than delaying visual learning. The following additional causes or interpretations of deficient visual performance have been suggested by various deprivation studies and are pertinent to the present results.

2. Passing a Critical Period

Visual deprivation may prevent the development of a visual response during an early *critical period*, when it would have appeared without learning, or would have been learned readily, whereas after that age a difficult process of learning or adjustment is required (Beach & Jaynes, 1954; Hess, 1962). What happens in the organism before the critical period, *increasing* the tendency for the response to occur, is usually called *maturation*. What happens after the critical period in animals kept in darkness, *decreasing* the tendency for the response to occur, could be described as *backward maturation*. This emphasizes the fact that *visual deprivation does not prevent visual development;* it alters the course of the development from what it would be with normal experience. In some respects the altered development is opposite to normal development, so that the long-deprived animal is inferior to an animal at the critical age or even to a newborn animal. If it can ever reach an equal level of visual performance, it needs a longer period of development under normal conditions to do so. In the monkey research, only those animals kept in darkness beyond the critical age of about 3 to 6 weeks failed to differentially fixate patterned or solid objects during the following several months, and only those animals showed a long-lasting deficit in visual discrimination. The critical period for avoidance of a visual cliff seemed to end around 1 month of age, whereas the critical period for learning to use vision to direct other behaviors did not end until 2 or 3 months of age. But such a description of the results does not clarify the *process* underlying the change in behavior after the critical age, which might involve any of the effects of deprivation described below.

3. Neural Deterioration

Visual stimulation is essential for development and maintenance of normal functioning of the visual system. While the periods of deprivation in the present research with monkeys were shorter than periods that have caused degeneration of neural structures in chimpanzees (Chow *et al.*, 1957), biochemical or other changes have been noted in various species after several months of visual deprivation which may or may not indicate functional impairment (Riesen, 1960, 1961a). Furthermore, individual cells of the visual cortex of kittens deprived of vision in one eye for 2 months or longer are unresponsive to patterned stimulation of the deprived eye (Wiesel & Hubel, 1963), while cortical cells of kittens raised with or without visual stimulation for 20 days or less show selective responsiveness to specific moving or stationary visual patterns similar to that found in adult cats (Hubel & Wiesel, 1963). Similar disruption in primates of this unlearned neural pattern selectivity, presumably underlying form perception and selective attention, would explain the effects of visual deprivation on visual pattern selectivity and form discrimination. Neural deterioration cannot now be excluded as a possible factor in any visual deprivation study.

4. OCULOMOTOR ABNORMALITIES

Spontaneous nystagmus was present in the longer-deprived infant monkeys, as in other deprived animals (Riesen, 1958). These jerky or rhythmic eye movements, along with binocular incoordination, undoubtedly hindered visual explorations and form discriminations by making it difficult to fixate or follow objects or to resolve fine patterns.

5. EXCESSIVE AROUSAL

Intense, complex, or novel visual stimulation of an animal reared under restricted sensory conditions may cause excessive neural or behavioral arousal as shown by hyperexcitability, emotional disturbance, stereotyped movements, random explorations, or failure to habituate to novel stimuli (Melzack, in press; Riesen, 1961b). Hebb (1955) has suggested that high levels of arousal interfere with the discrimination and selection of relevant cues from the environment. Animals reared in isolation as well as those reared with complete visual deprivation show excessive arousal. Both have difficulty in simple pattern discrimination, which may be attributable in part to "inability to select relevant patterns from the total sensory input (because of the high level of arousal) rather than to absence of pattern perception *per se*" (Melzack, 1962, p. 979). While restricted rearing makes carnivores hyperexcitable, Melzack (in press) suggests that primates tend toward apathy under such conditions, but that either state can interfere with utilizing sensory information.

6. COMPETING HABITS

The long-deprived infant monkeys tended to rely on touch and smell more than the less-deprived ones. Several developed a peculiar mode of walking, moving the hands in an arc in front as if to feel the way, which persisted long after the deprivation. When presented with a nursing bottle or food objects on a tray, some made sweeping movements of the hand or else "sniffed around" rather than reaching directly for an object. Such habits may add to the difficulty deprived animals have in learning to recognize objects and places visually.

7. DISTURBANCE OF VISUAL PREFERENCES

Decreased visual preference for configurational stimuli and increased preference for nonconfigurational stimuli are among the behavioral effects of visual deprivation. These changes could result from one or more of the effects mentioned above: deterioration of the neural mechanism for form perception (Section III, A, 3), difficulty in fixating small details, following contours, and converging the eyes for binocular vision (Section III, A, 4), or excessive arousal by patterned stimulation (Section III, A, 5). In view of the importance of patterned stimulation for spatial orientation and object perception (Fantz, 1961; Gibson, 1958), loss of visual interest in patterns could be responsible for the initial lack of visual capacity of primates reared in darkness as well as their later difficulty in visual learning. A possible mechanism for such effects is described in the next section.

B. Selective Attention and Perceptual Development

The differential-fixation method was used at the beginning of the present research program as the most feasible way of testing visual discrimination capacities of infant primates. But the consistent visual preferences that were discovered are of interest in their own right, having implications for the development of visual perception.

Some visual selectivity is essential for *visual exploration* of the environment, preventing random and chaotic eye movements on one hand and unvarying reflex fixations on the other. Selective visual exploration causes *differential exposure* to parts of the environment, and thus differential opportunities for learning. The particular visual preferences determine what the opportunities for learning will be. It would be clearly maladaptive for an animal to look predominantly at homogeneous surfaces or open space rather than at objects or patterned surfaces. In contrast, the visual interest in patterned surfaces and complex objects often shown by infant primates during the early months facilitates learning about the environment, since configurational variables provide most of the informational content of visual stimulation.

Animals have receptive capacities for an almost unlimited number of aspects of a variegated environment. What aspects are perceived at a given moment depends largely on what object or surface is fixated and which stimulus variables are attended to. So differential fixation and selective attention *are direct determinants of perception*. Experiences that change the selectivity change what is perceived; such changes may be the core of what is referred to as *perceptual learning*. This explanation of perceptual learning is an extension of the "stimulus differentiation" of Gibson and Gibson (1955): increasing selectivity in what is attended to will give increasing stimulus differentiation. However, instead of necessarily resulting in finer discriminations and more-veridical perceptions, perceptual learning in the present view may have varying effects. Under "normal" developmental conditions, behaviorally-important stimulus variables tend to be increasingly selected to give veridical perceptions and adaptive responses; but under other conditions, such as during and following visual deprivation, the changes may be toward greater selection of gross, uninformative, or irrelevant stimulus characteristics.

The proposed developmental role of selective visual attention may be clarified by a hypothetical analysis of the development of object perception, based in part on the results from infant monkeys. The following "phases" of development may overlap in time.

1. Certain types of pattern and differences in pattern are looked at selectively from birth because of the stimulating effect of the resulting patterns of neural impulses upon the central nervous system. Further maturation of the visual system, including more acute vision, a higher optimal level of arousal, and increased neural organization, cause an increase in the degree of interest in pattern complexity during the early weeks of life. This process is probably influenced by visual stimulation in a nonspecific way, as through practice in oculomotor coordination.

2. Early unrestricted experience in a variegated environment permits visual exploration and visual learning. The explorations are directed at first by the unlearned preferences. What is perceived and what is learned are influenced just as surely by this self-produced differential exposure as by the nature of the environment. *Objects* (as distinct from patterned or colored surfaces and empty space) are often set apart from the total visual input by compactness, high figure-ground contrast, binocular and movement parallax, and perhaps movement of the object. So objects tend to be favored in visual explorations, increasing the opportunity to differentiate and associate their distinguishing features, such as form, complexity, and solidity.

3. With the advent of manipulative and locomotor exploration, the possibilities for differential experience increase considerably. Selective visual attention now leads to selective grasping, mouthing, approaching, and climbing responses. But visual selectivity is, in turn, altered by the consequences of these responses, including, of course, reinforcing experiences from edible or noxious objects. Probably a more basic perceptual learning process involves feedback from physical contact, manipulation, and locomotion. The prior tendency to visually select objects facilitates this process by increasing the opportunity to associate

visual object characteristics with nonvisual ones. If this experience is varied enough, the animal will eventually generalize from the specific visual characteristics of familiar objects to the general object characteristics that are verified by touch or that support active exploration and play —e.g., "palpable form" characteristics, including solidity, orientation, texture, and the form of edges and surfaces. Attention will shift away from stimuli that have not been associated with tactile or kinesthetic differences during exploration and play. Thus the color, brightness, and patterning of smooth surfaces will tend to be ignored except when these aspects have served as cues for object recognition or spatial orientation, as do the color of food objects, shadows denoting solid objects, or changes in pattern density indicating depth.

4. These basic perceptual-learning processes may be supplemented by secondary types of learning under particular environmental circumstances. In some cases this will go counter to the original learning. The human child, for example, must learn to fixate details of patterned surfaces in order to recognize photographs and to read, while the laboratory primate may be given discrimination training. In the discrimination apparatus, a monkey will initially tend to select objects on the basis of their palpable form and can therefore easily be trained to visually discriminate objects differing in palpable form. In order to discriminate nonpalpable visual dimensions, the animal must go beyond what it has learned in everyday life and attend to stimulus features it has ignored.

This gives a reasonable explanation for the striking ease with which visually-experienced monkeys learned to discriminate objects differing tactually as well as visually, and for the close-to-chance responses to stimuli differing only in pattern, color, or brightness (Section II, D, 3), even when these had been discriminated during the early months of life in the visual preference tests (Section II, D, 1). Other results from monkeys and other primates can be interpreted in a similar way. Multidimensional or "junk" objects have repeatedly been found to be more quickly discriminated than unidimensional ones. This may be due not only to the presence of multiple cues, but to the invariable inclusion of palpable form differences. Menzel et al. (1960) presented six single-cue discrimination problems to two chimpanzees reared in isolation. The chimpanzees performed best on two problems involving palpable form differences—horizontal versus vertical and planometric versus stereometric. The horizontal-versus-vertical difference was discriminated almost as well as objects differing in many ways. Discrimination of flat plaques, differing only in the color, outline, or size of painted patterns, has been found to be more difficult than discrimination of multidimensional objects (Harlow, 1945; Warren & Harlow, 1952; see also Chapter 1 by Meyer et al. in Volume I). The supposed superiority of color over form as a discrimination cue may be due to varying form in restricted ways, often using what Gibson (1951) has called "disembodied varieties of form" (i.e., painted patterns, geometrical outlines, or pictorial representations of real objects). For primates that are experimentally naive and untrained in geometry, such forms would generally be inconsequential until their laboratory livelihood depended on discriminating them. When only the patterns on flat surfaces or the outlines of simple geometrical forms are varied in discrimination experiments, and other aspects of form that are more important for the behavior of primates under most natural conditions are omitted, the results may give a misleading impression of the importance of color, brightness, and size in the development of visual perception and recognition. This may also tend to perpetuate the erroneous belief that form and pattern are secondary and acquired stimulus variables. Further research is needed with widely-varying types of form problem to test this explanation of the discrepancy between the ease of form discrimination shown in the present experiment and the difficulty shown in previous studies of monkeys.

IV. CONCLUSION

The contribution of the research described in this chapter can be separated into empirical, methodological, and theoretical components. First, in accord with other types of evidence—neurological findings (Section III, A, 3), instinctive and learned form discrimination by newly-hatched birds (Section I, B, 2), unlearned depth discrimination by various animals (Section II, D, 2, a), form-discrimination learning by neonatal monkeys (Chapter 11 by Zimmermann and Torrey), pattern discrimination by visually-deprived chimpanzees in spite of incidental effects of deprivation (Section I, B, 1), and the noticing of form differences by congenital cataract patients after operation in spite of lack of a lens and other disturbing factors (Section I, B, 3, c)—the results described in Section II prove that the ability to resolve, organize, and discriminate visual configurations is present at birth or after postnatal maturation. These various lines of evidence also imply that experience and learning have their main effects on behavioral processes other than the reception and discrimination of patterned stimulation. Such processes may include the development of attention, cognition, and learning ability, and the learning of specific responses to what is seen. Second, the differential-fixation method makes it possible to study the young primate's perceptual model of the environment in an objective way, without either having to trace back from complex behavior sequences or having to project forward from the nature of the receptors and sensory input. Third, the results emphasize an often-neglected part of the perceptual process—selective attention. This is an active organismic determinant of what is looked at, what is perceived, and what is learned from experience, which interacts with the hereditary, structural determinants (i.e., sensory capacity), and the environmental determinants (i.e., sensory stimulation and the consequences of behavior).

Normal developmental changes in selective attention increase the correspondence between the perceived characteristics of the environment and those physical characteristics which are consequential for behavior. Abnormal changes, including the loss of unlearned interest in pattern, the failure to develop other interests in pattern or solidity, and the increased selection on the basis of nonconfigurational variables, are likely causes of the visual deficit shown by pattern-deprived animals in this and other studies. Because of these and other deteriorating effects of deprivation, it is now clear that deprived animals cannot substitute for newborn animals as subjects for the study of early perceptual development. Visual deprivation remains a useful technique, as in studying the importance of visual stimulation for neural and oculomotor function. It is also an indirect way of altering visual interests as a means of studying their role in perceptual development. Further knowledge of how experience enters into normal development, however, must come from the use of less extreme deprivation conditions, which vary the opportunity for specific types of perceptual learning.

If what is attended to determines what is perceived at various stages of development, as hypothesized in this chapter, new light is thrown on the nativism-empiricism controversy. Visual selectivity is clearly present from birth; however, the selected stimulus variables change with experience until in the mature primate the initial basis of selection may be largely obscured. In Section I, B, 3, various possible resolutions were given of the conflict between evidence that perception is innate and evidence that it is learned; these resolutions were based on the use of birds or primates as subjects, on the study of normal visual development or development following a period of deprivation, and on different operational definitions of perception. If perception is indicated neither by a particular instinctive response nor by a particular trained response, but by selective attention, whether unlearned or learned, then the nativism-empiricism controversy can also be resolved in terms of age of the subject: *perception is innate in the neonate but largely learned in the adult!* This is presented partly as a resolution which is as good as can be found, and partly to point out that no real solution is possible. It is perhaps best to be content with determining the various developmental factors that influence various perceptual processes and behaviors in subjects at various stages of phylogenetic and ontogenetic development, and to give up the attempt to prove either nativism or empiricism.

REFERENCES

Beach, F. A. (1960). Experimental investigations of species-specific behavior. *Amer. Psychologist* **15**, 1.

Beach, F. A., & Jaynes, J. (1954). Effects of early experience upon the behavior of animals. *Psychol. Bull.* **51**, 239.

Chow, K. L., & Nissen, H. W. (1955). Interocular transfer of learning in visually naive and experienced chimpanzees. *J. comp. physiol. Psychol.* **48**, 229.

Chow, K. L., Riesen, A. H., & Newell, F. W. (1957). Degeneration of retinal ganglion cells in infant chimpanzees reared in darkness. *J. comp. Neurol.* **107**, 27.

Fantz, R. L. (1954). Object preferences and pattern vision in newly hatched chicks. Unpublished doctoral dissertation, University of Chicago.

Fantz, R. L. (1956). A method for studying early visual development. *Percept. mot. Skills* **6**, 13.

Fantz, R. L. (1957). Form preferences in newly hatched chicks. *J. comp. physiol. Psychol.* **50**, 422.

Fantz, R. L. (1958a). Depth discrimination in dark-hatched chicks. *Percept. mot. Skills* **8**, 47.

Fantz, R. L. (1958b). Visual discrimination in a neonate chimpanzee. *Percept. mot. Skills* **8**, 59.

Fantz, R. L. (1958c). Pattern vision in young infants. *Psychol. Rec.* **8**, 43.

Fantz, R. L. (1961). The origin of form perception. *Sci. Amer.* **204** (5), 66.

Fantz, R. L. (1963). Pattern vision in newborn infants. *Science* **140**, 296.

Fantz, R. L. (in press). Pattern discrimination and selective attention as determinants of perceptual development from birth. *In* "Perceptual Development in Children" (Aline H. Kidd & Jeanne L. Rivoire, eds.). International Universities Press, New York.

Fantz, R. L., Ordy, J. M., & Udelf, M. S. (1962). Maturation of pattern vision in infants during the first six months. *J. comp. physiol. Psychol.* **55**, 907.

Gibson, J. J. (1951). What is a form? *Psychol. Rev.* **58**, 403.

Gibson, J. J. (1958). Visually controlled locomotion and visual orientation in animals. *Brit. J. Psychol.* **49**, 182.

Gibson, J. J., & Gibson, Eleanor J. (1955). Perceptual learning: differentiation or enrichment? *Psychol. Rev.* **62**, 32.

Grant, D. A. (1947). Additional tables of the probability of "runs" of correct responses in learning and problem solving. *Psychol. Bull.* **44**, 276.

Harlow, H. F. (1945). Studies in discrimination learning by monkeys: V. Initial performance by experimentally naive monkeys on stimulus-object and pattern discrimination. *J. genet. Psychol.* **33**, 3.

Harlow, H. F., & Zimmermann, R. R. (1959). Affectional responses in the infant monkey. *Science* **130**, 421.

Hayek, F. A. (1952). "The Sensory Order." Univer. Chicago Press, Chicago, Illinois.

Hebb, D. O. (1949). "The Organization of Behavior." Wiley, New York.

Hebb, D. O. (1955). Drives and the c.n.s. (conceptual nervous system). *Psychol. Rev.* **62**, 243.

Held, R., & Hein, A. (1963). Movement-produced stimulation in the development of visually guided behavior. *J. comp. physiol. Psychol.* **56**, 872.

Hess, E. H. (1956). Space perception in the chick. *Sci. Amer.* **195**, 71.

Hess, E. H. (1962). Imprinting and the "critical period" concept. *In* "Roots of Behavior" (E. L. Bliss, ed.), pp. 254–263. Harper, New York.

Hubel, D. H., & Wiesel, T. N. (1963). Receptive fields of cells in striate cortex of very young, visually inexperienced kittens. *J. Neurophysiol.* **26**, 994.

Jacobsen, C. F., Jacobsen, Marion M., & Yoshioka, J. G. (1932). Development of an infant chimpanzee during her first year. *Comp. Psychol. Monogr.* **9**, No. 1 (Whole No. 41).

Lehrman, D. S. (1953). A critique of Konrad Lorenz's theory of instinctive behavior. *Quart. Rev. Biol.* **28**, 337.

Lorenz, K. (1957). The nature of instinct. *In* "Instinctive Behavior" (Claire H. Schiller, ed.), pp. 129–175. International Universities Press, New York.

Melzack, R. (1962). Effects of early perceptual restriction on simple visual discrimination. *Science* **137**, 978.

Melzack, R. (in press). Effects of early experience on behavior: experimental and conceptual considerations. *In* "Psychopathology of Perception" (P. H. Hoch & J. Zubin, eds.). Grune & Stratton, New York.

Menzel, E. W., Jr., Davenport, R. K., Jr., & Rogers, C. M. (1960). Utilization of various visual cues by young isolation-reared chimpanzees. Paper read at Southeast. Psychol. Ass., Atlanta, Georgia.

Ordy, J. M., Massopust, L. C., Jr., & Wolin, L. R. (1962). Postnatal development of the retina, electroretinogram, and acuity in the rhesus monkey. *Exp. Neurol.* **5**, 364.

Pastore, N. (1958). Form perception and size constancy in the duckling. *J. Psychol.* **45**, 259.

Pastore, N. (1962). Perceptual functioning in the duckling. *J. Psychol.* **54**, 293.

Peterson, Marjorie E., & Rumbaugh, D. M. (1963). Role of object-contact cues in learning-set formation in squirrel monkeys. *Percept. mot. Skills* **16**, 3.

Portmann, A. (1962). Cerebralisation und ontogenese. *In* "Medizinische Grundlagenforschung" (K. F. Bauer, ed.), pp. 1–62. Georg Thieme Verlag, Stuttgart.

Pratt, C. C. (1950). The role of past experience in visual perception. *J. Psychol.* **30**, 85.

Riesen, A. H. (1951). Post-partum development of behavior. *Chicago Med. School Quart.* **13**, 17.

Riesen, A. H. (1958). Plasticity of behavior: psychological aspects. *In* "Biological and Biochemical Bases of Behavior" (H. F. Harlow & C. N. Woolsey, eds.), pp. 425–450. Univer. Wisconsin Press, Madison, Wisconsin.

Riesen, A. H. (1960). Effects of stimulus deprivation on the development and atrophy of the visual sensory system. *Amer. J. Orthopsychiatry* **30**, 23.

Riesen, A. H. (1961a). Stimulation as a requirement for growth and function in behavioral development. *In* "Functions of Varied Experience" (D. W. Fiske & S. R. Maddi, eds.), pp. 57–80. Dorsey Press, Homewood, Illinois.

Riesen, A. H. (1961b). Excessive arousal effects of stimulation after early sensory deprivation. *In* "Sensory Deprivation" (P. Solomon, P. E. Kubzansky, P. H. Leiderman, J. H. Mendelson, R. Trumbull, & D. Wexler, eds.), pp. 34–40. Harvard Univer. Press, Cambridge, Massachusetts.

Rosenblum, L. A., & Cross, H. A. (1963). Performance of neonatal monkeys in the visual cliff-situation. *Amer. J. Psychol.* **76**, 318.

Thorpe, W. H. (1956). "Learning and Instinct in Animals." Methuen, London.

Tinbergen, N. (1951). "The Study of Instinct." Clarendon Press, Oxford.

Tinbergen, N., & Kuenen, D. J. (1957). Feeding behavior in young thrushes. *In* "Instinctive Behavior" (Claire H. Schiller, ed.), pp. 209–238. International Universities Press, New York.

Tinbergen, N., & Perdeck, A. C. (1950). On the stimulus situation releasing the begging response in the newly hatched herring gull chick (*Larus argentatus argentatus* Pont.). *Behaviour* **3**, 1.

Tucker, Arlene F. (1957). The effect of early light and form deprivation on the visual behavior of the chicken. Unpublished doctoral dissertation, University of Chicago.

von Senden, M. (1960). "Space and Sight." Methuen, London.

Walk, R. D., & Gibson, Eleanor J. (1961). A comparative and analytical study of visual depth perception. *Psychol. Monogr.* **75,** No. 15 (Whole No. 519).

Warren, J. M., & Harlow, H. F. (1952). Learned discrimination performance by monkeys after prolonged postoperative recovery from large cortical lesions. *J. comp. physiol. Psychol.* **45,** 119.

Wertheimer, M. (1951). Hebb and Senden on the role of learning in perception. *Amer. J. Psychol.* **44,** 133.

Wiesel, T. N., & Hubel, D. H. (1963). Single-cell responses in striate cortex of kittens deprived of vision in one eye. *J. Neurophysiol.* **26,** 1017.

Zuckerman, C. B., & Rock, I. (1957). A reappraisal of the roles of past experience and innate organizing processes in visual perception. *Psychol. Bull.* **54,** 269.

Chapter 11

Ontogeny of Learning[1]

Robert R. Zimmermann and Charles C. Torrey[2]

Department of Psychology, Cornell University, Ithaca, New York

I. INTRODUCTION

A. Comparative and Developmental Psychology of Learning

When psychologists study learning, they usually ask one or more of these apparently straightforward questions: (1) Capacity: (a) What types of behavior can the organism acquire? (b) What is the asymptotic level of performance on a given task? (2) Rate of learning: (a) What is the initial level of performance? (b) How rapidly does the organism approach its asymptote of performance? (c) In what way does the

[1] Preparation of this chapter and the studies conducted at Cornell University were supported by Grants M-4516 and M-4355A from the National Institutes of Health, U. S. Public Health Service. Other previously unpublished experiments were supported by Grant M-772 from the National Institutes of Health while the senior author was a graduate student and postdoctoral research associate at the University of Wisconsin.
[2] Now at Lawrence College, Appleton, Wisconsin.

organism approach the asymptotic level of performance, i.e., what is the shape of the learning curve? (3) Transfer and generalization: What are the effects of prior learning on subsequent learning, usually tested in terms of: (a) stimulus generalization, (b) response generalization, (c) transfer to problems generated according to certain rules? (4) Retention: (a) Over how long an interval without practice can the organism maintain a previous level of performance? (b) In what manner does performance change as a function of time or activities intervening between original learning and later test? It is important to note that all of these are questions about the organism's performance, which is assumed to reflect a process, or several processes, referred to as "learning."

When these questions are asked about animals of different ages or species, it is usually assumed that all of the animals have had equal opportunities to learn. In other words, observed differences in performance are assumed not to result from differences in sensory or motor capacities or in motivation or reward. This assumption has been very important to comparative psychologists, and in particular to students of the phylogeny of learning in nonhuman primates (Harlow, 1958; Köhler, 1925; Warden, 1927), because their primary aim has been to evaluate the evolution of intelligence. With the theory of evolution as a model emphasizing continuity of structure and function, learning theorists have often speculated about evolutionary continuity in the learning process. But performance is also affected by other processes, such as emotion and motivation, so investigators of learning have tried to control these processes of secondary interest by selecting experimental conditions that were presumably equivalent for the various organisms being studied.

Bitterman (1960) has described this traditional method as *control by equation* and remarked that the prospects for successful treatment of interspecies differences through this method are "slim indeed." But he feels that a truly comparative psychology of learning might be developed instead on the basis of *control by systematic variation.* The main feature of this type of control is that functional relations which appear to differ for two species shall be studied at several levels of "contextual variables" such as deprivation of food or effortfulness of response. A corollary of this approach is that we are restricted to comparisons between species for one of which a reasonably well-developed theory of behavior has been worked out. Few of us would dispute Bitterman's nomination of the rat as the animal at present best qualified to play this role of standard species, but many animal psychologists would hope that valid comparisons need not, even in the beginning, be rat-relative.

Because of the large number of variables that may affect performance of a certain type of organism on a given task, the first steps of a comparative investigation should probably take the form of Bitterman's control by systematic variation. But then, having established optimal levels of contextual variables, systematic variation may not be necessary each time a difference is discovered. Simple classical and operant conditioning generally fail to reveal phylogenetic differences (see Chapter 7 by Warren in Volume I), so comparative studies usually involve tasks requiring differential responses to the environment. The experimenter can therefore design a method that is optimal for the particular response that will later serve as the indicator of learning. Then, using the parameters that maximize performance, the questions listed above can be attacked.

Ontogenetic comparisons of learning, of course, involve all of the problems of phylogenetic comparisons. But studying the ontogeny of behavior also brings unique problems. Developmental psychologists have traditionally tried to find variables controlling behavioral changes that can be plotted against age. One analytic approach to the problem of development is to divide the continuous elaboration of behavior from birth to maturity into relatively discrete stages or periods. Recently, students of learning have been particularly interested in the notion of "critical periods" for learning, for infantile stimulation, and for the formation of social relationships. In a survey of critical periods in a wide variety of species, Scott (1962) suggested that the maturation of emotional and motivational processes is crucially important in determining what, and how rapidly, an animal will learn at a particular stage of development. Thus, affective behavior, already a source of difficulties for the comparative psychologist, enters the picture again in the study of development. The critical-period hypothesis points out that the performance of any animal in a learning situation must be evaluated in terms of what that animal has learned during earlier critical periods, and that task-related motivational and emotional variables must be controlled for animals of differing ages. Failure to consider both of these problems could lead to erroneous conclusions about ontogenetic changes in learning ability.

A final problem in studying the ontogeny of learning in primates arises from the traditional methods of measuring learning. The demonstration of instrumental learning presupposes that the young organism has enough motor coordination to produce a reliable effect on the organism-environment relationship. The vast majority of instrumental learning tasks have rested on locomotive and manipulative skills that develop relatively slowly and are limited or rudimentary in the infant monkey, ape, and human. Along with the need to develop tasks suitable for the infant primate, the student of learning must also identify potential stimuli that are reinforcing for the young animal, and then must invent ways to deliver such stimulation.

In the following sections the theoretical and practical questions brought out here will be discussed in detail when specific circumstances arise. In general, our position is that motivational and procedural problems must definitely be dealt with, but are obstacles to analyzing the evolution and ontogeny of learning.

B. Observational Literature

Although the experimental literature on the ontogeny of primate learning has centered on the study of the rhesus monkey (*Macaca mulatta*) (see Section I, C), there remains a considerable body of observational data on general and selected aspects of development in other species. Most of this material is widely scattered throughout the zoological and older psychological literature. But we felt it important to examine this literature, both to discover what sorts of information were available and to seek a broader base for assertions about the ontogeny of primate learning than was offered by purely experimental studies.

Field studies are largely devoted to taxonomic and ecological problems. Behavior, treated in terms of the general "habits" of the species, arboreal or ground-dwelling, nocturnal or diurnal, etc., is usually subordinated to detailed descriptions of morphology, diet, and diseases. Observations of young animals center on adult-infant relations, the manner in which the young are transported, the duration of nursing, and the development of social behavior (see Chapter 15 by Jay). While many apparently learned behaviors, such as eating and drinking habits, cooperation, detour behavior, and the use of objects, receive passing mention, their ontogenesis remains a mystery. Where behavior is a central topic, it is the organization and structure of the social group that has been most extensively studied. Reports on the behavior of captive animals, and particularly of births in captivity, are occasionally more informative, but for most species births in captivity are rare, and attention tends to center on obstetrical information, problems of maintenance, and ability.

A second class of publications includes a variety of observational studies of general and selected aspects of development (Foley, 1934; Hines, 1942; Jacobsen *et al.*, 1932; Jensen, 1961; Kellogg & Kellogg, 1933; Knobloch & Pasamanick, 1959a, 1959b; Richter, 1931; Riesen & Kinder, 1952), many of which have implicit or explicit conceptual links to the Gesell scales for human development. That is, they emphasize the development of the most general sensorimotor integrations: prehension, posture, and locomotion. Relatively little attention is given to the environmental context in which such integrations develop and are expressed. Thus, while these studies may suggest further lines of research on the comparative and developmental aspects of learning, they do not in themselves provide much information on such aspects.

The observational data, fragmentary and scattered though they are, provide a picture of general primate development which, in its broad outlines, holds well across most primate families. Following birth and maternal cleaning of the young, the infant remains quite literally attached to the mother for hours or days, its mouth to her breast even while not actively nursing. Shortly, however, the infant begins to take note of its visual environment, releasing the nipple to look around, or making tentative reaching movements with one hand while still clinging to the mother. In some species, alternatives to the usual ventroventral orientation of infant to mother are found. In the marmoset (*Callithrix jacchus*), the infant's favored companion is frequently an adult male rather than the mother (Fitzgerald, 1935; Lucas *et al.*, 1927). The picture of the infant primate as a well-developed clasping mechanism with strong tendencies toward ventral contact with soft, furry surfaces, is by and large accurate for all species.

· The beginnings of independent behavior are marked by brief excursions in the immediate vicinity of the mother, with frequent retreats by the infant or retrievals by the adult. During a period ranging from a month to half a year or more, this independence increases to the point where the infant spends as much time with his age-mates as with an adult. Play behavior appears and becomes elaborated during this period, and continues thereafter through adolescence. In both howler monkeys (*Alouatta palliata*) and gibbons (*Hylobates lar*), Carpenter (1934, 1940) observed play sequences which, with the coming of maturity, merged more or less continuously into adult patterns of dominance and submission. Mason (1961b) indicated the importance of learning in this transformation. The young animal, through play and social interaction with his fellows, and the reinforcements contingent on reaction to social stimuli, quickly learns to avoid aggressive animals and to dominate submissive animals.

Adult-young relations also suggest the role of learning in the achievement of general and social adjustment in primates. Yerkes and Tomilin (1935) cited observations in support of the view that the mother chimpanzee (*Pan*) provides active tutelage of the infant in the development of locomotor behavior. Fitzgerald (1935) reported similar observations in the case of the marmoset. In several species, the transition from liquid to solid diet appears to be largely under maternal control, the mother at first discouraging and later encouraging oral experimentation with solids (Nolte, 1955; Yerkes & Tomilin, 1935). Fitzgerald suggested that the lapping of liquids from a dish is learned by the infant marmoset through observation of the adult. Other instances of imitational learning, particularly in the elaboration of gymnastic play, have been described by Yerkes and Tomilin (1935).

Arboreal species offer interesting instances of adult-young cooperation in some ways analogous to *umweg* behavior. The adult howler, for example, often forms a bridge with its body to assist the young animal in crossing from one tree to another. Carpenter (1935) observed similar behavior in the Panamanian red spider monkey (*Ateles geoffroyi*). In other species, adults carry otherwise-independent juveniles over difficult crossings, setting them down on the other side (Carpenter, 1935). Nolte (1955) described three-dimensional *umweg* solutions in immature bonnet macaques (*Macaca radiata*), the young animal solving difficult crossings by descending to the ground and ascending the neighboring tree to follow the band.

In a few instances, observational and experimental methods have been combined in the longitudinal study of single animals. The performance of the chimpanzee Viki on a series of imitation problems (Hayes & Hayes, 1952) is of interest in connection with the suggestions elsewhere of natural imitational learning by young primates. Kellogg and Kellogg (1933) compared the performance of the young chimpanzee Gua with that of their son Donald in a number of tasks such as bladder and bowel control, instrumentation problems, and *umweg*. As is well known, the comparisons favored the hypothesis of more rapid initial maturation of learning ability in the chimpanzee. These two studies indicate the chimpanzee's capacity to adjust to a unique social environment, but they do not indicate the behavior to be expected of animals either in the laboratory or in the wild.

In summary of this section, then, observational data help to evaluate the "ecological validity" of the standard laboratory tasks, and to identify, at least tentatively, some of the important variables controlling primate behavior. But naturalistic observation alone cannot take us very far in understanding the ontogeny of learning, for the technique necessarily confounds many developmental and phylogenetic variables.

C. Laboratory Studies

A survey of the literature on laboratory studies of learning in young primates indicated that the wide gaps in our knowledge preclude extensive comparative analysis at this stage. We are unable to locate any laboratory studies of learning in young prosimians or platyrrhine (New World) monkeys. The experiments with young apes are limited in number, in the range of tasks studied, and in rigor of design and execution. In comparison, the available information on the rhesus monkey is extensive and reliable.

This concentration on the rhesus macaque and relative neglect of other species springs from a number of considerations. The rhesus monkey's rate of sensorimotor development, its low mortality and high rate of reproduction in captivity, and its considerable intellectual capacities combine to recommend this animal as an ideal subject for the study of ontogeny (Harlow, 1959). By contrast, prosimians, even as adults, show great limitations in learning ability as indexed by standard tests (Jolly, 1964); platyrrhine monkeys, by and large, have poor breeding records in captivity (Hill, 1960, 1962), and the great apes are expensive to maintain and their slow development makes the collection of longitudinal data a time-consuming process.

The classification of experiments presented in this chapter, while fairly traditional, also reflects the authors' notions of increasing task complexity. Because of the distribution of experiments found in the literature, our discussion will be concerned chiefly with the analysis of learning in the infant rhesus macaque. Although the available data from other species will be introduced under the appropriate headings, it is apparent from the outset that generalizations about the ontogeny of primate learning can only be tentative, pending more thorough study of a variety of species.

II. CLASSICAL CONDITIONING

Mason and Harlow (1958a), training infant rhesus macaques from the third day of life, found unequivocal conditioning of motor activity as measured by a stabilimeter and by independent observers. Tone and shock were paired daily for eight trials and tone was presented alone for two test trials per day. Forty reinforced trials sufficed to establish a conditioned response. Presenting the tone in other situations yielded no evidence of generalization. Few conditioned responses were made after a 15-day period of no practice.

Solenkova and Nikitina (1960) trained monkeys of various species (two hamadryas baboons [Papio hamadryas], three baboon hybrids, and one macaque hybrid) and of ages ranging from 14 to 109 days. The response indices were leg flexion and respiratory changes. The rate of formation and consolidation of conditioned responses to a tone paired with shock increased with age. Both response components were unstable in younger monkeys, and the prominence of respiratory responses declined with age. Some motor responses remained after 90 nonreinforced trials in the only animal run on extinction.

Green (1962) extended Mason and Harlow's stabilimeter technique, studying conditioning, extinction, generalization, and retention in rhesus monkeys aged 1, 30, and 300 days at the beginning of training. The conditioned stimulus was a 1,000-cps tone and the unconditioned stimulus was shock delivered through the grid floor of the stabilimeter cage. Observation was again used to supplement the stabilimeter records. Two indices of conditioning were derived from the observations: *increased activity* (leaping, hopping) and *decreased activity* (crouching, freezing). These measures followed different courses as a function of age. All groups were alike in showing an initial steep increase in activity, but after the third day of training this component declined in the youngest group. While there were no over-all differences in the decreased-activity measure of conditioning, the curve for the 1-day-old group showed a rather abrupt rise following the fifth day of training. A combined measure based on both observational indices showed highly similar development of conditioned responses for all groups throughout the course of training. In later extinction, generalization, and retention tests, the youngest group showed the greatest deficits in performance.

Thus, three studies have demonstrated classical conditioning in very young monkeys. Evidence that rate of conditioning changes with age is equivocal, but appears to depend on choice of index. Green's study and that of the Russian investigators suggest that age has more profound effects on the form of response than on the rate of learning.

III. LOCOMOTION AND MANIPULATION

While the demonstration of classical conditioning makes relatively simple demands on the organism in terms of response differentiation, instrumental tasks usually require a more highly organized behavioral repertoire. As was pointed out in Section I, A, reliable evidence of instrumental learning rests on the organism's ability to produce reliable changes in its relation to the environment. Although a given change, as of location, can be effected by a variety of response-means, such as swimming, walking, running, or flying, each of these response sequences is likely to show greater organization and differentiation than the type of diffuse reaction which is susceptible to conditioning by the Pavlovian method. In a sense, it is only with instrumental situations that the notion of "task" enters the learning picture. Much that goes on under the instrumental paradigm as applied to adult organisms can be thought of as "learning what the task is" rather than as the establishment of particular sequences of response to a defined stimulus pattern.

Mason and Harlow (1958c) reported on the development of approach responses of infant rhesus monkeys to a nursing booth in the animal's home cage. At the beginning of each of the 12 daily feedings, the animal was placed at the opposite end of the cage and was allowed 30 seconds to enter the booth. Over the period of the experiment, which covered the third through the thirty-second day of life, the infants rapidly improved in approach either to an unlighted booth or to a booth illuminated with a flashing green light. A third group of animals showed reliably less improvement when a 5-minute delay of feeding followed entrance into a lighted booth. The green light was not preferred in another situation where the monkeys were not fed.

The development of reaching and transporting solid objects to the mouth apparently begins at least as early as the twentieth day of life in the rhesus monkey. Mason and Harlow (1959) presented an array of uncolored pieces of banana, of brightly and variously colored pieces of banana, or of heterogeneous foods, for 10 minutes each day beginning when 30 rhesus monkeys were 20 days old. Records were kept of hand contacts, mouth contacts, and consumption of food. The infants tested with the colored banana and mixed foods showed reliably higher contact scores than those tested with plain banana. Yet during all 40 days of testing, only 19 of the 30 monkeys ate at least one-half piece of food, a finding which supports the view that manipulation can be reinforcing in and of itself (see Chapter 13 by Butler). Informal observations at our own laboratory have shown that solids placed in the home cage are tactually and orally explored with increasing frequency, and that by 50 days of age all of the infants are consuming some portion of foods presented in this manner.

That manipulative behavior appears early in the life of the infant rhesus monkey was further demonstrated in a study by Harlow *et al.* (1956). The responsiveness of six infant monkeys to mechanical devices was tested starting at 16 to 36 days of age. Manipulation appeared spontaneously on the introduction of each device and increased with age and experience. In most cases, successful responses or discriminations indicative of learning increased with practice. These results suggested that manipulative responsiveness probably existed before any formal testing was begun. A study by Mason (1961a) supported this conclusion, suggesting that manipulative responsiveness of the young rhesus monkey is limited only by its motor ability (see Chapter 13 by Butler). Unfortunately, this potential source of reinforcement has not been exploited more fully in the analysis of learning in the infant rhesus monkey.

IV. SINGLE-PROBLEM DISCRIMINATION LEARNING

A. Spatial

Using a single-unit Y maze, 15- and 45-day-old rhesus monkeys were given two trials a day in a spatial discrimination task (Mason & Harlow, 1958b). The reward was return to the individual's living cage. Both groups mastered this problem rapidly, taking about six to nine trials to reach a criterion of learning. After learning the appropriate response, both groups were tested on a reversal of the position habit. No significant differences were found between age groups. Harlow (1959) reported extreme emotionality in the early phases of reversal learning.

B. Nonspatial

1. Apparatus and Procedures

The Wisconsin General Test Apparatus (WGTA; see Chapter 1 by Meyer *et al.* in Volume I) has been used in some studies of the development of learning, particularly those using infant monkeys over 2 months old. For younger monkeys, different techniques have been devised. For example, Zimmermann (1961) substituted two adjacent nursing booths for the one used by Mason and Harlow (1958b), so that neonatal rhesus monkeys could be trained on nonspatial discriminations in the monkey's living cage. The booths were movable and left-right positions were reversed randomly. Experiments using this apparatus have established that the infant rhesus monkey can learn to discriminate differences of brightness, pattern, and form before it is 1 month old (Sections IV, B, 2 and 3).

While the home-cage apparatus, providing a maximum amount of freedom during the testing sessions, was adequate for extremely young infants, it did not control variables that are often controlled in visual-discrimination experiments. Such variables include the position of the animal at the time the stimuli are exposed and the length of time between exposure of the stimuli and the opportunity to respond.

In order to control these variables, a Grice-type discrimination apparatus used extensively with rats was modified for use with neonatal and infant monkeys (Zimmermann, 1961). The subject was placed in a start box with a transparent door separating it from a choice chamber (Fig. 1). An opaque door separated the subject visually from the discriminanda inserted in the rear of the divided goal area. When the opaque door was raised, the subject was usually delayed in the start box for 2 seconds, after which the transparent door was raised and the subject was permitted to approach the goal area. The discriminanda were painted or mounted on boards and a hole was drilled in the center of each stimulus to permit the insertion of a nursing-bottle nipple. The reward was always 5 to 12 cm³ of infant-monkey formula (Blomquist & Harlow, 1961). About 90% of the monkeys born at the Wisconsin laboratory could be trained to locomote over a smooth wooden floor from the start box to the rear of the goal chamber before 10 days of age. At the Cornell laboratory, with the floor of the apparatus covered with a soft cotton rug, reliable approach responses appeared in most neonatal monkeys before 5 days of age.

Techniques for training monkeys in this apparatus were developed with 60- to 90-day-old infants. These studies indicated that young monkeys would run for three to six trials at each of their scheduled 4-hour feedings. Since then, we have found that time between feedings can be reduced to about 2 hours if no supplementary feedings are given until all testing is completed, but that five trials per feeding is probably optimal. Discrimination performance of extremely young animals varies greatly from day to day, so the criterion of learning requires consistent performance for more than 1 day. Our standard criterion of learning is 21 correct of 25 choices on each of 2 consecutive days of training.

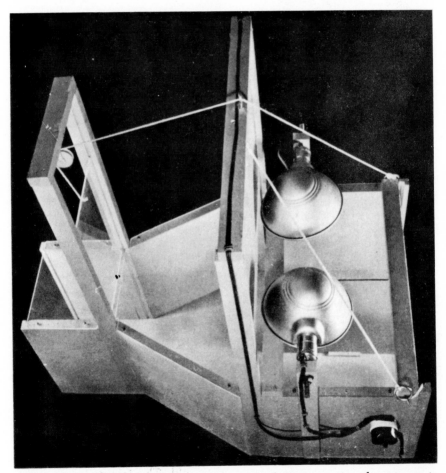

FIG. 1. Discrimination apparatus with doors raised. During testing, the entrance to the start box (left) is closed by means of a sliding door, and the stimulus board is slid into the back of the goal chamber (right). (From Zimmermann, 1961.)

Fɪɢ. 2. Black-white discrimination booths during test session. (From Zimmermann, 1961.)

2. Brightness Discrimination

Zimmermann (1961, Exp. 1) trained neonatal rhesus monkeys on a black-white discrimination using the home-cage apparatus (Fig. 2). The infants were trained to climb up a black or white ramp for the reward of infant-monkey formula. The ramp, the entire feeding booth, and even the caps of the nursing bottles were matched in color. Figure 3 shows the development of learning by a group of four rhesus monkeys that began training on the first day of life and a second group of four infants that were trained with identical gray booths for the first 10 days of life. The infants trained from the first day of life were making 90% correct responses by the time they were 11 days old. The infants trained with identical gray booths for the first 10 days took only 2 days to approach this same level of performance on the black-white discrimination.

After 20 days of training for each group, a reversal problem was introduced (Zimmermann, 1958). The previously positive booth was now made negative and vice versa. This reversal condition continued for 10 days and was followed by 5 days of retraining to the original stimulus, which in turn was followed by 5 days of training under the reversal condition. The results of this study were strikingly similar to those described above for the spatial reversal problem. All the infants were highly emotional and in many instances refused to drink the formula offered as reward for a correct response. No reliable effects of age were found. The

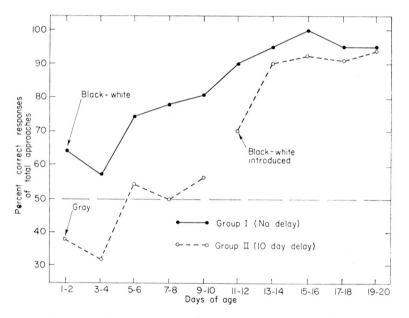

Fig. 3. Black-white discrimination performance of two groups of rhesus monkeys. (After Zimmermann, 1961.)

data indicate that the neonatal monkey is capable of learning a reversal problem in 5 days (about 50 trials) of training. Although all animals showed some emotional disturbance in the first reversal session, the animals reversing to black had to be force-fed in the test situation. The effects of reversing to black as opposed to reversing to white are quite evident in Fig. 4. Through the three changes in stimulus conditions, all of the infants rewarded for approaching white reached 88% correct responses before any infant rewarded for approaching black. This is the earliest known example of an influence of stimulus preference on discrimination-learning performance of infant primates.

Four of the infants from the above study continued in a series of seven repeated reversals with these same stimuli. Each condition was in effect for 5 consecutive days. Progressive improvement over repeated reversals is indicated in Fig. 5, which shows the percentage of correct responses on the first day of each 5-day reversal, including the data from the previous experiment. The curves appear to show that learning to

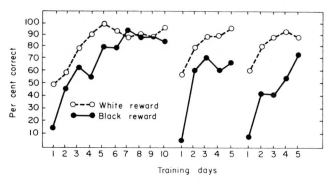

Fɪɢ. 4. Mean percentage of correct responses by monkeys fed on white and on black during reversal training.

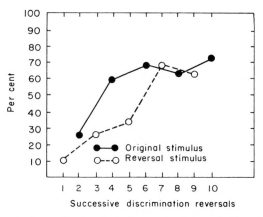

Fɪɢ. 5. Mean percentage of correct responses to the original and reversed problems on the first day of each reversal.

choose the original positive stimulus improved more rapidly than learning to choose the original negative stimulus, but this difference was not reliable. However, the reliable black-white difference noted above persisted throughout the entire series of reversals.

In another study (Zimmermann, 1963), four 1-year-old rhesus monkeys with extensive discrimination training and four naive adult rhesus monkeys were trained in a WGTA to discriminate between two gray surfaces differing in brightness. The stimuli were selected by having human observers in a simulated primate-testing situation make "same" or "different" judgments of stimuli from a gray scale of known reflectance. A stimulus pair was considered discriminable if no "same" judgment occurred in five paired comparisons. Seven pairs of stimuli were selected, the central pair being used as the training stimuli. Two monkeys in each group were rewarded for responses to the lighter gray and two for responses to the darker. After achieving a criterion of learning, tests of transfer to the other pairs of stimuli were begun. Two adults were culled for failure to learn the original problem in 60 days (1,500 trials). The remaining adults reached criterion after 17 and 26 days of training, while the four infant monkeys took a mean of 8.75 days. The results of the transfer tests appear in Fig. 6. Performance of both groups deterio-

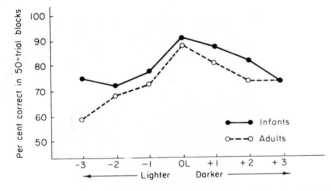

FIG. 6. Brightness-transposition performance of 1-year-old and adult rhesus monkeys.

rated as the transfer stimuli deviated from the original training stimuli. Also, both groups made reliably more errors to the lighter stimulus pairs than to the darker pairs. No reversal of transposition occurred; that is, the monkeys either responded in the direction of original learning or fell to chance performance.

3. PATTERN AND FORM DISCRIMINATION

Pattern and form discrimination by newborn rhesus monkeys were also studied using the home-cage apparatus (Zimmermann, 1961, Exp. 2). The differential cues were painted directly on white booths. In one problem horizontal and vertical stripes were painted on the ramp and walls, and in another problem an equilateral triangle and a circle were painted on removable pressed-board plaques inserted into the rear of the booth (Fig. 7). The procedures in this experiment were identical to those used in the black-white discrimination problem. After 6 days of age, performance on the pattern and form problems was consistently inferior to performance on the black-white problem (Fig. 8).

The Grice-type discrimination apparatus (Fig. 1) was first used to study the effects of age on the learning of a single form discrimination (Zimmermann, 1961, Exp. 3). Three groups of rhesus monkeys with little or no prior discrimination training were tested five times a day at 4-hour intervals, five trials per session, starting at 11, 29, and 61 days of age. The 11-day-old group was the slowest to learn (Fig. 9). Performance of

Fig. 7. Triangle-circle discrimination booth during test session. Under actual test conditions the nipples were not inserted until after a response was made. (From Zimmermann, 1961.)

the other two groups improved similarly for the first 5 days of training, but the 61-day-old group improved further on day 6 while the 29-day-old group did not. Statistical analysis revealed that the 11-day-old group was reliably inferior to the older groups, but that the older groups did not differ reliably. A group of eight infant monkeys with a mean age of 65 days, but with a history of discrimination of colored surfaces and objects, solved this same problem in a mean of less than 75 trials. It is apparent that both age and experience are important variables in the development of discrimination learning in young rhesus monkeys.

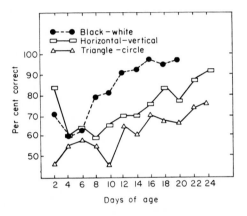

Fig. 8. Performance of neonatal rhesus monkeys on three discrimination problems. (After Zimmermann, 1961.)

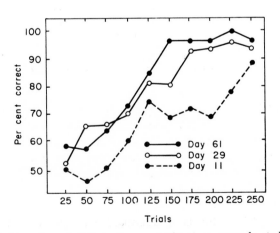

Fig. 9. Performance of infant rhesus monkeys on triangle-circle discrimination. (After Zimmermann, 1961.)

4. Relative Difficulty of Form and Brightness Discrimination

The inefficient learning of a form discrimination, both by the 11-day-old subjects in the Grice-type apparatus and by 1-day-old monkeys in the home-cage apparatus, contrasts sharply with the rapid learning by animals of comparable ages discriminating black from white in the home-cage-feeding-booth situation (Section IV, B, 2). The superior performance of the black-white groups might have resulted from the greater total area of differential cues present in the black-white stimulus array, or it might reflect some difference between the ability to discriminate brightness and the ability to discriminate form. To test these alternatives, black and white stimuli were made with the same total area as the triangle and circle. Four 11-day-old monkeys were tested by Zimmermann (1961, Exp. 4) in the Grice-type apparatus on the same schedule as described in Section IV, B, 3. Their performance is compared in Fig. 10 with that of

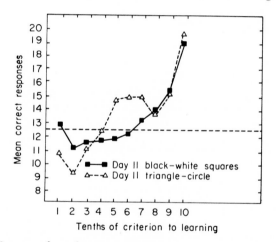

Fig. 10. Mean number of correct responses in Vincent tenths of training to criterion (21 correct in 25 responses) on brightness and form discriminations. (After Zimmermann, 1961.)

the 11-day-old triangle-circle group from the previous experiment. Each group required a mean of 193.7 trials (8.8 days) to reach the criterion of learning. Thus, the discrimination of form is not intrinsically more difficult than the discrimination of brightness when the stimuli are equated for a variable known to affect the learning performance of adult organisms.

5. Color Discrimination

a. Projected light. Color is a dominant stimulus dimension in discrimination learning by adult monkeys (see Chapter 1 by Meyer *et al.* in Volume I). To study the development of color discrimination in the neonatal monkey, two projectors were placed behind the goal chambers of the Grice-type apparatus so that they illuminated sheets of tinted Plexiglas in the rear of each chamber. This provided a homogeneous colored light in each goal chamber. Brightness and saturation were controlled by manipulating the number of sheets of Plexiglas inserted into the rear of each goal chamber. A rotating disc containing a continuous slit which varied in width, placed directly in front of the projectors, produced a continuously changing intensity at each stimulus surface. Under conditions of greatest saturation, the stimuli appeared deep green and bright amber to the human observer. Four groups of four naive infant rhesus monkeys began discrimination training at 10, 20, 30, and 40 days of age. Age had no reliable effect on learning; all groups required a mean of over 250 trials to solve this problem.

b. *Painted surfaces.* The results of the color-discrimination experiment described above were surprising in view of the color-discrimination ability of the adult monkey (Grether, 1939) and in view of the effects of color on the development of eating solid foods (Section III). Furthermore, Harlow reported in the 1959 Messenger Lectures at Cornell University that infants were discriminating between feeding and nonfeeding terry-cloth surrogate mothers differing in color before they were 20 days old. The inefficient learning found in the previous experiment may have been produced by the use of projected light, as opposed to light reflected from a surface, and by the random variation of brightness and saturation.

To test this possibility, four 10-day-old and four 20-day-old rhesus monkeys were trained to discriminate between painted surfaces. The stimuli were 8-inch by 10-inch Masonite boards painted green or orange. The differential cues covered the entire rear of the respective goal chambers of the apparatus. Using the standard five trials per feeding five times a day, the 10-day-old subjects took a mean of 4 days, or 100 trials, to reach the criterion of 21 correct responses out of 25 choices on each of two consecutive days, while the 20-day-old infants took a mean of 2 days, or 50 trials, to reach the same criterion. Thus, as was found for other dimensions, it appears that discrimination of colored surfaces by the infant monkey improves with age.

Eight naive adult monkeys were trained in a WGTA on a similar problem. The discriminanda were 5-inch green or orange squares mounted in stimulus holders that covered the foodwells and held the squares at an angle of about 60 degrees from the horizontal plane (Fig. 11). The adult animals took a mean of 5.5 days, or 137.5 trials, to reach the standard criterion. This result serves to remind us that different modes of testing are not always comparable. In order to expose the foodwell, the adult monkey in the WGTA had to push at or near the base of the stimulus holder. The lower edges of the colored squares were about 7/16 inch above the bottom of the base, so the adult monkeys may have pushed at places spatially separated from the colored surfaces (cf. Chapter 1 by Meyer *et al.* in Volume I). On the other hand, no such separation was present in the infant test situation. The learning scores of the best four adults were comparable to those of the 20-day-old infants.

Fig. 11. Monkey's-eye view of stimulus holders in Wisconsin General Test Apparatus.

c. Effects of reduced differential-cue area. The experiment on the relative difficulty of form and brightness discrimination suggested that a major variable in the efficiency of learning was the area of the discriminanda. To further evaluate this variable, the infant subjects from the experiment on the discrimination of painted colored surfaces were given a series of transfer tests in which the stimuli that had occupied 100% of the rear wall of the goal chamber were reduced to 50%, 25%, 12.5%, 6.25%, and 3.125% of the area of that wall. The adult subjects received comparable tests with reduced colored areas on the 5-inch square. In both cases the differential cues were centered on a gray background. The data presented in Fig. 12 support the traditional finding that performance progressively deteriorates as the area of the differential cues is reduced. The consistent inferiority of the adult animals may readily be attributed to the increased cue-response separation with the reduction of area. Spatial contiguity would have little effect upon infant performance since the approach of the head to the center of the differential cue did not change with changing area.

6. SIZE DISCRIMINATION AND TRANSPOSITION

The subjects from the above experiment (both infants and adults) were then trained on a size discrimination, in which one of the discriminanda occupied the central 25% and the other the central 12.5% of the total area of the stimulus holders. The infants were tested in the Grice apparatus and the adults in the WGTA. Under these conditions the infant monkeys, with a mean age of 27 days, required about 600 trials to learn. Some adult monkeys required fewer trials, but some failed to reach the criterion of learning in 60 days of training. Size transposition was then tested with stimuli that were 100% versus 50%, 50% versus 25%, 12.5% versus 6.25%, and 6.25% versus 3.125% of the total area of the holders.

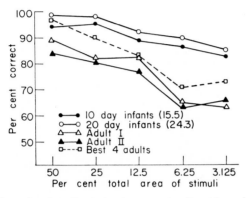

FIG. 12. Transfer of color discrimination to stimuli with reduced colored area. "Adult I" and "Adult II" are two groups of four monkeys, assigned randomly.

Both infant and adult monkeys showed consistent transfer only with the stimuli smaller than the training pair. This was true whether approach to the larger or smaller of the two stimuli had been rewarded in original training.

7. STIMULUS GENERALIZATION

Seventeen infant rhesus monkeys between 11 and 32 days of age were trained on one form discrimination (triangle versus circle, with the circle negative) until they met the standard criterion. They were then tested on 28 pairs of generalization stimuli illustrated in Fig. 13 (Zimmermann, 1962). Each test session consisted of two generalization trials with the same stimulus pair, separated by four training trials using the original triangle-circle stimuli. The order of presenting the generalization stimulus pairs was randomized for each animal, and during the generalization trials either response was rewarded with standard milk formula. The training trials provided information about retention, so that an infant could be retrained to criterion if performance of the originally learned response fell below 80% correct. The experiment was conducted daily until each stimulus pair had been presented 20 times to each animal. The percentage of responses by all infants to the left-hand stimulus of each

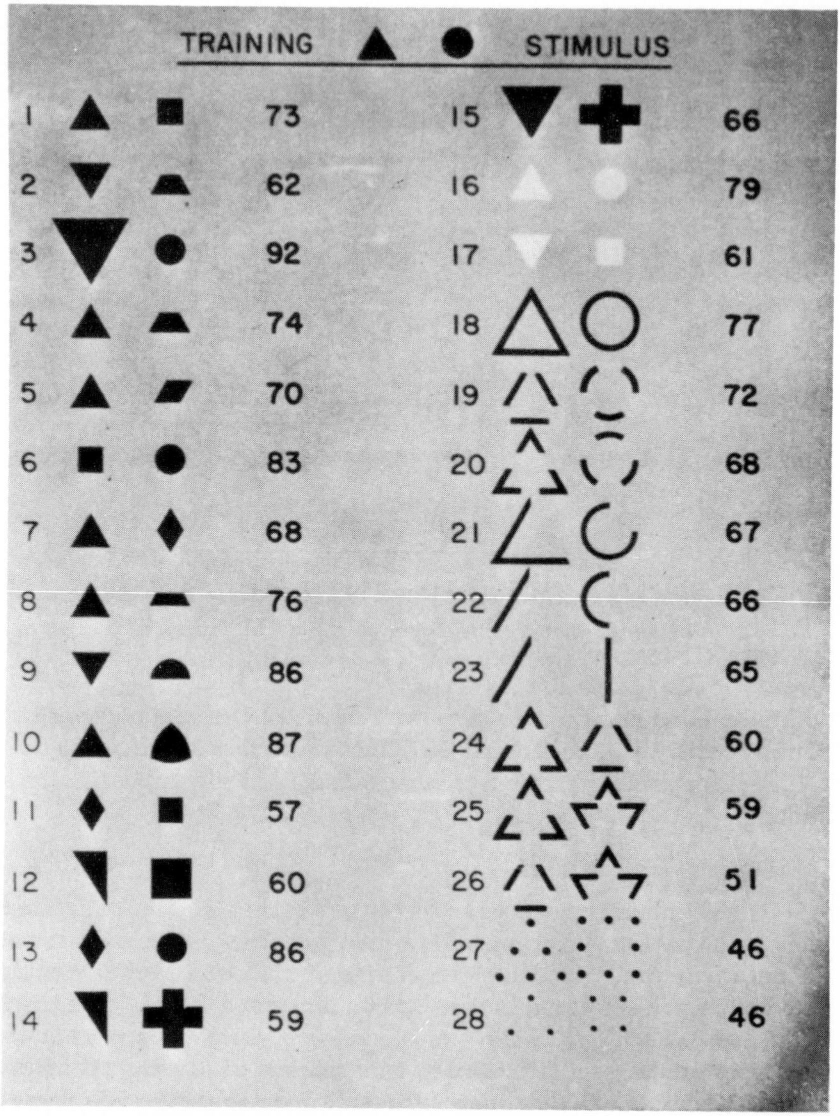

FIG. 13. The training pair and 28 generalization pairs. The percentage of choices of the left-hand stimulus is shown next to each generalization pair. (From Zimmermann, 1962.)

generalization pair appears in Fig. 13. The infant monkeys responded consistently to new pairs of form stimuli that appear to the human to resemble the original stimuli. Differential responding was maintained when both of the original stimuli were made brighter (pair 16) and also when they were outlined rather than filled (pair 18). About the same degree of generalization occurred to the sides of the triangle as to the angles (pairs 19 and 20). Stimulus-generalization scores approached chance when neither of the original stimuli was present, or when such changes as brightness, a new negative figure, and rotation of the triangle were compounded (pair 17).

In a study recently completed at the Cornell laboratory, eight infant rhesus monkeys with a mean age of 65 days learned a single triangle-circle discrimination and were then tested for *transfer of training* on four series of five stimulus pairs similar to those used in the previous study. This experiment differed from the above in three important ways: (1) only responses to the more triangular object were rewarded; (2) the original-training-stimulus pair was never reintroduced; and (3) each stimulus pair was presented in four-trial blocks five times a day, and series I to IV were presented to all animals in that order. The percentage of correct responses to each stimulus pair is shown in Fig. 14. A compari-

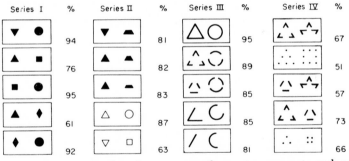

FIG. 14. Transfer stimuli with percentage of correct responses to each pair.

son of these results with those of the previous experiment reveals one consistent and striking trend. Those stimulus pairs to which reliable generalization occurred in the previous experiment were the ones showing reliable transfer effects in the present experiment. Similarly, chance or near-chance responding occurred to the same pairs in the two studies. The transfer technique tended to increase the amount of differential responding to those stimulus pairs to which the degree of stimulus generalization had been relatively high.

At the University of Wisconsin and in our own laboratory, adult monkeys have been tested with stimuli identical to those described in the above two experiments. Infant performance was as good as or better than adult performance and the degree of generalization to the various stimuli was almost identical. However, the performance of adults tested with the transfer technique improved reliably over repeated testing. This was not true of the performance of infant monkeys in either of the above experiments.

8. Discrimination of Stimuli Differing in More Than One Dimension

So far, this review of discrimination learning has discussed only the discrimination of stimuli differing in only one dimension (brightness, color, form, or size). The number of dimensions differentiating stimuli is known to affect rate of learning and degree of transfer by adult monkeys (see Chapter 1 by Meyer *et al.* in Volume I).

In a study recently completed at the Cornell laboratory, 5-, 10-, 20-, and 30-day-old experimentally naive infant rhesus monkeys were trained on a discrimination problem in which the stimuli differed in color, form, and size. The stimulus colors were deep blue and bright red; the forms were an equilateral triangle and a circle; the sizes were about 1 square inch and 7 square inches. Since only five animals were available in each age group, it was not possible to use all combinations of color, size, and form in the original learning problem. To control for possible preferences in color, size, or form, positive and negative stimuli were counterbalanced as closely as possible.

All infant rhesus monkeys born in our laboratory began pretraining in the Grice-type apparatus on the first day of life. The apparatus was identical to that described for the other neonatal experiments except that its floor was covered with a deep-pile gray cotton rug. All subjects were trained to the criterion of learning and then began tests of transfer to stimuli with the number of discriminable dimensions reduced. Each original training problem generated 18 reduced-cue pairs (Fig. 15), but, since the number of animals was limited, the six reduced-cue categories rather than the 18 stimulus pairs were presented equally often. The infants received 200 trials on each of the six reduced-cue categories of color-size, color-form, form-size, size, form, and color. Also, during this time, the infants received 200 trials on the original problem in order to control for possible interference effects. The total of 1,400 trials followed the same schedule as the original training (five trials per feeding), with five stimulus pairs presented each day.

Six additional animals with a mean age of 85 days, and with previous experience on single-dimension discriminations of color, size, and form, were also tested in the Grice-type apparatus. For three animals the positive stimulus on each dimension had been positive in previous training;

for the other three, the positive stimulus on each dimension had been negative.

A group of adult rhesus monkeys was adapted to a WGTA and trained to push 5-inch-square sheet-metal stimulus holders that were tilted at an angle of 60 degrees from the horizontal plane and attached to hardboard bases which covered the foodwells (Fig. 16). Stimuli identical to

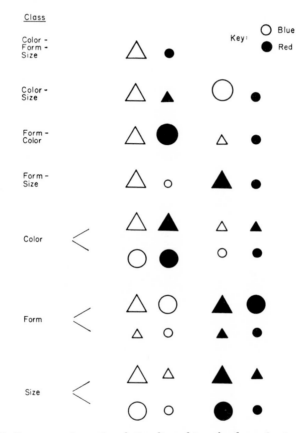

FIG. 15. Representative series of stimuli used in color-form-size transfer tests.

those described for the infants were painted on 5-inch-square pieces of ¼-inch hardboard that could be inserted into the stimulus holders. The adult subjects were trained to the same criterion as the infants and then given the same number of trials per category on the transfer test.

Learning curves for the original problems are shown in Fig. 17. In computing the percentages for this figure a score of 23, the mean of the criterional learning score and the maximal daily score, was inserted after a subject had achieved the criterion of learning. The mean number of days required to reach criterion was 4.75, 5.00, 2.00, and 1.75 for the 5-, 10-, 20-, and 30-day-old animals, respectively. The 85-day-old positive-transfer animals showed almost 100% transfer to achieve the criterion of learning, while the negative-transfer animals required a mean of 2 days, or 50 trials. But the adult animals required a mean of 9.5 days. The inferior performance of the adults was probably due to spatial separation of cue and response (cf. Chapter 1 by Meyer *et al.* in Volume I). Adult monkeys, tested with similar cues presented as objects rather than as painted patterns, learned this discrimination as rapidly as 30-day-old infants. Another important point is that multidimensional cues facilitate

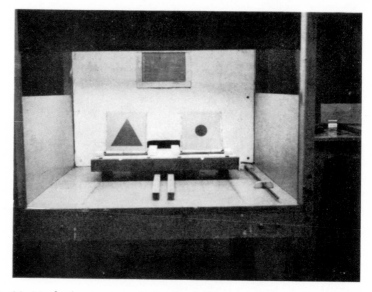

Fig. 16. Monkey's-eye view of stimulus holders containing representative stimuli for color-form-size transfer tests in WGTA.

learning. The 5- and 10-day-old infants in this study required less than 5 days of training, on the average, to reach criterion, while 11-day-old animals trained on a single dimension (brightness or form) took a mean of 8.8 days to reach the same level of learning (Section IV, B, 4). The performance of 20- and 30-day-old animals in this experiment was similarly superior to that of monkeys of comparable age tested on single-dimension discrimination problems.

Performance on the transfer tests showed significant effects of age and stimulus dimensions. Although there were some differences in the effects of age depending on the stimulus dimension (Table I), the interaction was not significant. The 5-day-old group was consistently inferior on all transfer tests including responses to the original problem during transfer of training. The stimuli fall into three homogeneous classes: (1) Original learning, color-size, and color-form; (2) form-size and color; (3) form and size. The dominance of the color dimension and the additivity of form and size cues are consistent with the results of other experiments (see Chapter 1 by Meyer *et al.* in Volume I). The most important finding of this experiment is that infant monkeys less than 1 month old respond to changes in the number of stimulus dimensions in almost the same manner as mature adults.

Using more traditional primate-testing methods, including solid foods as rewards and random object pairs as stimuli in the WGTA, Harlow *et al.* (1960) found progressive improvement with age in initial object-quality-discrimination-learning performance by infant monkeys ranging

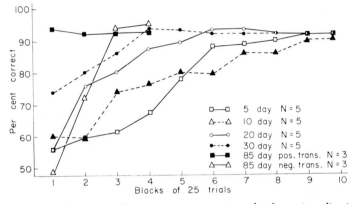

FIG. 17. Mean percentage of correct responses in color-form-size discrimination training.

TABLE I

PERCENTAGE OF CORRECT RESPONSES IN 200 TRIALS OF TRANSFER TESTS

| Class | Age (Days) | | | | | | |
	5	10	20	30	85	Adult	Mean
Color-Form-Size	90	93	92	96	92	96	93
Color-Size	88	91	90	95	91	93	91
Color-Form	89	91	89	96	95	94	92
Form-Size	79	95	79	88	90	87	85
Color	76	83	84	90	93	86	85
Form	69	74	66	78	80	75	74
Size	69	74	68	78	80	78	75
Mean	76	83	84	90	89	86	85

in age from 60 to 150 days. The 150-day-old infants performed as well as 360-day-old animals. The mean number of errors before 10 consecutive correct responses is shown in Fig. 18 for the five age groups used by Harlow *et al.* (1960) and for the 5-, 10-, 20- and 30-day-old infants of the Cornell experiment just described. The 20- and 30-day-old infants of the latter experiment were as good as or superior to the 60- and 90-day-old infants of the former experiment.

The source of the discrepancy between the two experiments may lie in procedural variables. The similarities in test procedures of the two experiments were as follows: 25 trials were run daily, correct and incorrect positions were determined by use of a Gellermann sequence, and

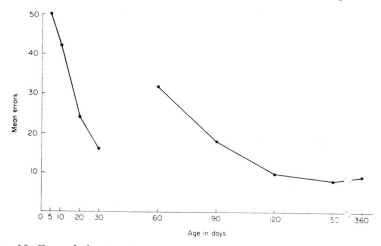

FIG. 18. Errors before attaining criterion on initial discrimination problem. Ages 60–360 days from Harlow *et al.* (1960).

noncorrection technique was used throughout. The differences were: Harlow *et al.* tested in the WGTA, used solid food reward, and gave 25 consecutive trials, 5 days a week, with stimuli that might induce stimulus perseveration (preference or avoidance). In our experiment the infants were tested in the Grice apparatus for milk formula reward (their only source of nourishment) in blocks of five trials spaced throughout the day, 7 days a week, with stimuli that apparently did not induce any consistent stimulus perseveration. Any or all of these variables could have contributed to the differences in performance.

9. REVERSAL AND NONREVERSAL SHIFTS

Tighe (1964) trained infant rhesus monkeys, 108 and 138 days old, in the Grice-type discrimination apparatus, and 15-month-old and adult rhesus monkeys in the WGTA on a reversal-nonreversal-shift problem

(Kendler & Kendler, 1962). All of the subjects had had previous discrimination-learning experience. The stimuli were flat or raised objects
painted with horizontal or vertical stripes. Half of the monkeys were
rewarded for responding consistently to flat versus raised and the remaining half for responding to horizontal versus vertical. Thus, in all cases, one
relevant and one irrelevant dimension were present during the original
learning of the discrimination. After reaching the criterion of learning,
half of the animals were rewarded for responding to the previously irrelevant dimension, a nonreversal shift, while the others were rewarded
for reversal responses to the previously relevant dimension. In all cases,
whether in original learning, reversal, or nonreversal, the flat-versus-
raised discrimination was learned more rapidly than the horizontal-
versus-vertical discrimination, and no reliable age differences were found.
Nonreversal shifting was easier than reversal learning for all age groups
and for both stimulus dimensions.

C. Summary

The studies of single-problem learning and transfer of the learned response indicate that the young rhesus monkey develops the ability to
discriminate traditional stimulus dimensions very early in life. More important, however, is the demonstration that primary stimulus generalization, transposition, or other forms of transfer, and sensitivity to such
variables as problem difficulty, stimulus preference, and stimulus dominance do not change radically after the first few months of life. In fact,
in four experiments it was demonstrated that the effects of these variables
change very little after 1 month of life.

V. LEARNING SETS

A. Discrimination

Data on the formation of discrimination-learning sets (see Chapter 2
by Miles in Volume I) have been reported for the immature rhesus monkey by Mason et al. (1956), by Harlow et al. (1960), and by Torrey
(1963). Mason et al. (1956) used six infants with prior experience in
manipulation tasks (Harlow et al., 1956; see Section III) and single discrimination problems. At about 165 days of age a series of 616 six-trial
problems was begun. Concurrently, the animals were trained on delayed-
response and patterned-strings problems. The formation of a learning set
was shown by orderly and progressive improvement over problems in
performance on trial 2, although the learning set was formed much more
slowly than in normal adults (see Chapter 2 by Miles in Volume I).

In a more detailed analysis of the relation of age and formation of
learning set, five groups of animals, about 90, 120, 150, 180, and 390 days
old, were given several hundred six-trial discrimination problems by Harlow *et al.* (1960). As in the earlier study, the monkeys were concurrently
trained on delayed-response and patterned-strings problems (see Section
VI). There was no indication of learning-set formation as measured by
trial 2 performance for animals in the 90- and 120-day groups (Fig. 19).
Animals ranging from 150 to 390 days old at the beginning of training
gradually and progressively improved. Under these conditions, 1-year-old
infant monkeys had not reached the adult level of proficiency in the formation of learning sets.

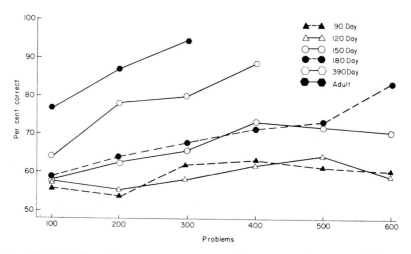

Fɪɢ. 19. Trial 2 performance as a measure of learning-set formation. (After Harlow *et al.*, 1960.)

In an experiment designed to evaluate the effects of degree of learning
and number of trials per problem on the rate of learning-set formation,
Torrey (1963) trained eight animals on 400 object-discrimination problems, beginning when the monkeys were about 255 days old. Four of
the animals were run on each problem to a criterion of 10 consecutive
errorless trials. The other four were run on a shifting schedule of 25 trials
per problem for the first 25 problems, 20 trials per problem for problems
26 through 50, and 15 trials per problem thereafter. As it turned out, the
groups received about the same number of trials per problem. Trial 2
performance improved progressively, and when Torrey's data are plotted
against problems they agree reasonably with comparable data of Harlow
et al. (1960).

Learning-set formation has also been studied in young chimpanzees, gorillas (*Gorilla gorilla*), and orangutans (*Pongo pygmaeus*). Hayes *et al.* (1953) gave a series of discrimination problems to eight chimpanzees ranging in age from 15 months to 26 years. A criterion for the learning of individual problems was used, and correction of errors was permitted. Hayes *et al.* concluded that "capacity to acquire a learning set does not increase with age, after the point where training becomes practical" (Hayes *et al.*, 1953, p. 103). The data were offered as preliminary, and it seems likely, in view of the clear age effects for rhesus monkeys and for humans (Reese, 1963), that better-controlled work would reveal an age gradient in chimpanzees as well.

Rumbaugh and Rice (1962) presented 500 six-trial problems to a 3.5-year-old gorilla, a 3-year-old chimpanzee, and a 4.5-year-old orangutan. Performance on trial 2 of the last 100 problems was 70% correct for the gorilla, 60% for the chimpanzee, and 55% for the orangutan. These performance levels are astonishingly low, even allowing for the uncertainty about what these ages mean in terms of general development of apes. A more detailed analysis of motivational and procedural variables is necessary in these species.

B. Oddity

Ten rhesus monkeys were tested on 256 two-position oddity problems (see Chapter 5 by French in Volume I) at 20 months and again at 36 months of age (Harlow, 1959). Performance on trial 1 improved with practice at both ages, the initial level for the second series being about equal to the terminal level for the first. It is evident that these 2- to 3-year-old animals were inferior to adults. However, because of the test-retest design, the evidence for improvement over the year between test series remains equivocal, and a study of separate groups at different ages is desirable.

In a second study, also reported by Harlow (1959), clear evidence for interproblem improvement in oddity performance was shown in animals as young as 12 months. These animals were also retested at 36 months, after half of them had had practice on discrimination-learning set and the others on delayed response. Negative transfer from discrimination-learning set to oddity-learning set was revealed (cf. Chapter 3 by Levine in Volume I).

VI. OTHER FORMS OF INSTRUMENTAL LEARNING

A. Delayed Response

The ontogeny of the ability to solve the delayed-response problem (see Chapter 4 by Fletcher in Volume I) is described by Harlow *et al.* (1960). Four groups of 10 rhesus monkeys began a 0-second and 5-second delayed-response test in the WGTA at 60, 90, 120, and 150 days of age. After 18 weeks of testing, with 900 trials at each delay, 12 additional weeks of testing at delay intervals of 5, 10, 20, and 40 seconds were begun. All groups performed at or near chance level during the first 100 trials of 5-second delay, after which the performance of the older animals improved more rapidly than that of the two younger groups. The 120- and 150-day-old monkeys reached asymptotic performance (about 90% correct) after 600 trials. The 60-day-old infants never attained this level of performance, and the 90-day-old infants only approximated this level after 900 trials. Performance levels at 0-second delays were 5 to 10% higher for all groups. Comparison of the 90- and 150-day-old groups shows that age is more important than experience in the development of this ability. Both of these groups approached asymptotic performance at 225 days of age, even though the older group had had 300 fewer trials. Also, these young infants required 600 to 900 trials to reach a level of performance that adult monkeys achieve in 200 trials. At longer delays, the performance of the 60- and 90-day-old animals was consistently inferior to the performance of the 120- and 150-day-old animals and a control group of adults. It was concluded that the infant monkey is not capable of performing efficiently at longer delays until 8 or 9 months of age, while 5 months appears to be a minimal age for attaining a high performance level on the shorter delays.

B. Hamilton Perseverance Test

In the Hamilton perseverance test as adapted for monkeys, the subject is faced with four identical boxes with lids that close after being opened. One of the four boxes contains food, and the subject's problem is to find the food in the least number of moves. From trial to trial the reward is shifted with the one stipulation that the same box never contains food on two consecutive trials. Errors are usually defined as repeated responses to the same empty container during any one trial. Optimum solution of this problem may be described as "avoid the box that last contained food, then respond in a sequence." Another efficient mode of attack is to respond consistently in a particular sequence.

Harlow (1959) reported a study in which three groups of rhesus monkeys, 12, 30, and 50 months old, were tested in the four-container situation. The 50-month-old monkeys made fewest errors from the start and approached an asymptote of almost no errors after only 240 trials. The 12-month group made almost three times the number of errors in the initial trials, and at the end of 600 trials both the 12-month group and the 30-month group were making about three errors in 60 trials. Only the 50-month group rapidly adopted a systematic or sequential-response approach to this problem. For example, at the end of 200 trials, these animals changed from one sequence of response to another with a mean frequency of only 10 times in 60 trials. Even after 600 trials, the mean number of sequence changes for the younger two groups did not go below 20 per 60 trials.

C. Reward-Directed Tasks

Simple discrimination problems, multiple-sign problems, and learning-set tasks have formed the backbone of primate learning research for the past 20 years. The popularity of these tasks can be attributed to the denumerability of response alternatives, together with the presumed initial indifference of the discriminanda. These features have lent plausibility to theoretical schemata categorizing such tasks or ordering them by complexity, primarily on the basis of logical considerations (Bitterman, 1962; Harlow, 1951). There is, however, an older tradition in primate research, dating back at least to Köhler's (1925) work with chimpanzees. In such problems, the reward is not suddenly revealed when the animal makes the correct response, but is visible from the outset and can be expected to exert a strong influence on the course of behavior from the very beginning. The animal's actions may be conceived as *reward-directed* rather than *cue-directed*.

We believe that fundamental differences in problem-solving behavior may well exist, depending on whether or not the reward is visible and influential in determining the course of behavior. Thus, we are not entirely happy with discrimination-learning set as a model for the development of insightful behavior in *all* types of problems, though we concede its applicability to cue-directed behaviors. It is clear that over a long series of similarly-constructed discrimination problems, primates gradually shift from seemingly "trial-and-error" learning to seemingly "insightful" learning (see Chapter 2 by Miles in Volume I). We should be wary, however, of concluding that the ontogeny of problem solution in reward-directed situations is similarly gradual and continuous. It seems to us advisable to reserve judgment for as long as possible on an issue of such fundamental historical importance.

For these reasons we treat patterned-string problems, bent-wire problems, instrumentation and *umweg* problems as a group distinct from the more carefully analyzed instrumental tasks. In this respect, our taxonomy of learning tasks differs from that of Bitterman (1962), who has classified "insight" problems as a special type of Thorndikian task, the purpose of which is to investigate "productive" rather than "reproductive" solutions. We prefer to retain, for the time being, a distinction between cue-directed and reward-directed tasks.

1. PATTERNED STRINGS

In a patterned-strings problem, two or more lengths of chain or string extend away from the subject in some pattern similar to those shown in Fig. 20, with food attached to the far end of one string. Mason and Harlow (1961) reported a series of developmental experiments on patterned-strings performance of rhesus monkeys. The most important of these experiments, from our viewpoint, was the first, using animals in seven age groups. The infant groups were 60, 90, 120, 150, and 360 days old at the beginning of training. For comparative purposes, groups of adolescents and adults were also trained on similar problems. Three string patterns were presented for 400 trials each. For half the subjects in each group, the order of presentation was parallel-crossed-pseudo-

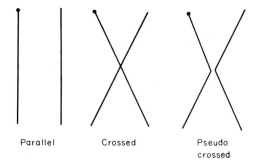

Parallel   Crossed   Pseudo
                               crossed

FIG. 20. Patterned-strings problems used by Mason and Harlow (1961).

crossed; for the others, the order was crossed-parallel-pseudocrossed.

Certain general trends are clear: All groups improved with practice on all problems, except when the initial level of performance left no room for further improvement. For all ages, the crossed-string pattern was more difficult than the parallel, whether or not the animals had had prior training on the other problem. In general, older animals learned more rapidly and achieved higher terminal performance than younger animals, regardless of problem and regardless of prior experience on other problems.

In addition to these major findings, there were a number of interesting relations between age and pattern of difficulty and between age and transfer effects: Mason and Harlow suggested that young monkeys tend to reach directly toward the food and hence to select the string originating nearest the food. The difficulty of the crossed-strings pattern for young animals, and the relative ease with which they handle the pseudocrossed pattern, support the notion that young animals are dominated by proximity. By contrast, older animals tend to display pattern-dominated behavior at first, although they are more flexible about adopting new modes of response than are the infants. Both adolescent and mature animals had initial difficulty with the pseudocrossed pattern, regardless of the immediately prior problem. However, these animals rapidly improved and reached high terminal performance levels on the pseudocrossed problem.

The pseudocrossed pattern appears to be of crucial importance. More could have been learned had animals of varying ages been given this as an initial problem with subsequent transfer to the crossed or parallel problem. As it is, Mason and Harlow ran seven adolescent animals on the sequence parallel-crossed, and six on the sequence pseudocrossed-crossed. Performance on the parallel-crossed sequence is very similar to that for the comparable age group in the first experiment. Comparison of performance on an *initial* pseudocrossed problem with corresponding data for the same problem as a transfer task shows that the early below-chance performance in the first experiment was due to the prior training. It is also clear that for adolescents, and presumably for adults as well, the crossed and pseudocrossed problems interfere strongly with each other, regardless of presentation order. It would be interesting to know whether order of presentation would be equally unimportant in the case of the pseudocrossed and parallel problems. We would predict that prior training on the former would produce *no* initial deficit on the latter.

So far as we have been able to discover, there have been no correspondingly detailed studies of the development of patterned-strings performance in other species. Riesen *et al.* (1953) gave numerous string problems to three gorillas, aged 8 months, 2 years, and 2.5 years. Conclusions were tentative, because of the small number of animals, but it was suggested that the infant gorillas did almost as well as adult chimpanzees tested by Finch (1941), better than 4- and 5-year-old chimpanzees tested by Birch (1945), and better than adult monkeys.

2. Instrumentation

Other studies have reported performance of young and adult chimpanzees on tasks comparable to the patterned-strings problems. Schiller (1952) found a maturational gradient in the solution of instrumentation problems by chimpanzees. A series of problems proceeded in increasing order of difficulty from a string baited at the near end to be drawn in through the cage bars, through rakes and straight sticks to be used singly and in combination to obtain food outside the animal's cage. Older animals were increasingly successful with later problems in the series. Five infants, 1 to 2 years old, were able to solve problems only so long as the tool did not have to be adjusted in relation to the food. Six animals, 3 to 4 years old, solved these problems more rapidly and were able to master simple adjustment of the tool to the food, but could not solve problems when the stick was presented wholly inside the cage. Eight animals, 5 to 8 years old, were much quicker in solving problems, and successfully used a short stick to reach a longer one which in turn was used to reach the food.

The improvement in problem solving with age was related to changes in manipulation of sticks in a nonproblem situation. The youngest animals typically handled a stick passively, without relation to other objects, and spent relatively little time manipulating it. Chimpanzees ranging from 2 to 4 years old used a stick to poke at other objects. Older animals, when given two sticks, one of which had a hollow end, tended to join the sticks spontaneously within the first few minutes. It is noteworthy, however, that the introduction of food disrupted this behavior of the older animals. Thus, one chimpanzee that had previously joined two sticks within the initial 5 minutes of spontaneous play took a total of 46 minutes to join and use the sticks in the presence of food.

3. BENT-WIRE PROBLEMS

Davis *et al.* (1957) compared the performance of preadolescent chimpanzees, adult chimpanzees, and adolescent rhesus monkeys on a series of 40 different bent-wire problems. Each problem consisted of a short piece of heavy-gauge wire with one or more right-angle turns, on which was threaded a candy Life Saver. The wire was held in a clamp and the animal's task was to remove the candy within a limited time. Sixteen rhesus monkeys, 18 to 24 months old, were given 13 trials on each problem. Two adult wild-born chimpanzees and five laboratory-born chimpanzees between the ages of 5 and 7 years were given 12 trials on each problem.

On all problems, the performance of young chimpanzees was intermediate to that of the monkeys and the adult chimpanzees. Errors increased with the number of straight wire segments, and problems involving ingoing terminal segments or segments directed away from the subject were more difficult than problems involving outgoing terminal segments or segments directed toward the subject. No analysis of interaction between these pattern variables and age or species was reported.

Davis *et al.* concluded that the developmental changes in performance on these problems reflect general improvement of perceptual organization in the direction of "increased complexity and stability" as a function of experience accumulated over the years. Interpretation of the species differences is complicated by deprivation conditions. The apes were fed three times a day, while the monkeys were fed only once at the end of the day, and hence were more deprived at the time of testing.

VII. OVERVIEW

In view of the limited data available on nonhuman primates other than the rhesus monkey, our summary will be restricted to a discussion of this organism.

The results of the studies described in Sections V and VI contrast markedly with those of studies of conditioning (Section II), locomotion and manipulation (Section III), and single-problem learning (Section IV). In the latter tasks, it appears that less than 30 days of maturity are necessary for infant rhesus monkeys to achieve a level of performance equal to that of adult monkeys. Although the rate of learning may be somewhat slower, the neonatal animal has the capacity to learn. Once the problem is learned, the infant's performance is affected by certain variables in the same way as the performance of the mature organism. Conversely, the studies in Sections V and VI reveal moderate (5 month) to rather enormous (4 year) delays in the development of certain types of problem-solving ability. It is also evident that extensive training does not aid in bringing monkeys up to high levels of performance. In these problems, developmental status appears to be the critical variable.

Considering the traditional questions posed in the introduction:

(1) Capacity. The survey reveals that three broad classes of problems may be identified according to criteria of above-chance performance or significant changes in responses defined by the experimenter:

(a) Those that can be solved at birth or shortly after birth, depending on the maturity of the organism (sensory-motor organization): classical conditioning, instrumental responses of locomotion and manipulation.

(b) Those that can be solved during the first month of life and depend on the sensory-motor organizations evident in (a) above: spatial, color, pattern, and object discrimination, and discrimination reversal.

(c) Those that can be solved only after several months of normal growth and whose solutions are not facilitated by extensive early training but develop primarily as a function of age: delayed response, perseverance, learning set, and patterned strings.

One other distinction separates classes (a) and (b) from class (c). Although young animals may require considerable training to reach a high level of performance in classes (a) and (b), their asymptotic performance is equal to that of mature organisms. In class (c), training which exceeds four to five times that required by older animals does not produce equal levels of performance. These ratios could undoubtedly be increased if it were not for the fact that the developing organism may reach the level of maturity necessary to achieve a high level of perform- ance before more trials or problems can be administered.

(2) Rate of learning. The question of rate of learning is partially answered in (1), but deserves special consideration for specific types of problems. In general, within the limits of the problem, more rapid learn- ing occurs with increased age. Part of this improvement may be attrib- uted to the development of sensory and motor systems, but other factors are obviously involved.

(a) Initial level of performance in discrimination tasks is usually manipulated by the experimenter through procedural methods so that group performance satisfies the acceptable chance level. However, cer- tain stimuli produce above- or below-chance performance scores during the initial phases of all testing. In the case of instrumental locomotion and manipulation, the initial level of performance is probably closely related to the maturation of sensory and motor systems in early phases of development and to emotional and motivational factors in later phases of development. Finally, changes in initial responses in the classical- conditioning tests surveyed may be complicated by the fact that there is a change in the mode of response to shock.

(b) The rate at which an infant monkey approaches asymptotic performance is a function not only of the class of problem (object dis- crimination, patterned strings, etc.), but also of more specific problem features. For example, the rate of learning an object or pattern dis- crimination increases through 120 to 150 days of age, but the rate of increase is not uniform for all dimensions. Our research indicates that brightness- and color-discrimination ability improves little after 20 to 30 days of age, while the ability to discriminate size, form, and depth appears to improve at least through 90 to 120 days of age. As one might expect from the data on single object-discrimination problems, the rate of learning a single object-discrimination-reversal problem also appears to reach a maximum at about 120 to 150 days of age. Other examples of important differences within a class of problems are found in patterned- strings tests, in which improvement is a function of the pattern, and in delayed response, in which the relation of performance to age is a func- tion of the length of delay.

(c) With the changing rate of learning, there is a corresponding change in the form of the group learning curves, typically from a posi- tively to a negatively accelerated curve. However, these group curves do not depict individual performance accurately, especially in the older groups where learning is characterized by several days at chance or near-chance performance (position habits) and then a sudden rise to criterion or near-criterion performance.

(3) Transfer. The performance of infant monkeys on transfer and stimulus-generalization tests falls into categories similar to those described under "Capacity."

(a) Transfer to single or altered dimensions, generalization, and transposition do not appear to change significantly after 1 month of age. In fact, the performance of neonatal monkeys has in several cases been superior to adult performance. Progressive intraproblem improvement in a repeated-reversal problem with stimuli constant is found in monkeys less than 2 months old.

(b) Interproblem transfer in learning-set tests, both object-quality discrimination and oddity, is considerably inferior to that of the mature animal even after 2 years of age. Although no data are available, we would guess that this is also true of reversal-learning set. Transfer of training in patterned-strings problems is apparently affected by the original pattern learned and the sequence of transfer patterns.

(4) Retention. Ontogenetic information about retention is available from only two kinds of task, classical conditioning and delayed response. If we can generalize from this small amount of information, it would appear that retention improves with age in the young monkey. This is one of the major areas of learning that has been almost completely neglected by scientists studying the ontogeny of learning in nonhuman primates.

Other classical paradigms and variables of learning do not appear in the literature. We have no information on the ontogeny of escape and avoidance learning and no information on the effects of deprivation and schedules of reinforcement on performance as a function of age. These gaps are, of course, understandable to those close to the neonatal primate. When faced with a situation requiring escape or avoidance of shock or other aversive stimuli, neonatal monkeys usually respond by "freezing," and studies of deprivation and schedules of reinforcement are precluded at the present time by the almost-heroic efforts and personal sacrifices which graduate students, employees, and sometimes professors make to maintain healthy neonatal monkeys.

From the developmental point of view, the data from traditional tests provide us with two broad generalizations. First, it is obvious that certain learning abilities appear extremely early in the life of the infant rhesus monkey. The ability to solve a problem may be present at 1 day of age, and within a matter of days performance may change from chance to 100%. Second, as Harlow (1959) has pointed out, after the initial burst of the development of certain learning abilities in early infancy, "the appearance of new learning powers" is characterized by a gradual or progressive improvement. There does not appear to be any sudden onset or waning of abilities at a specific period of life. In fact, within the range of ages studied (1 day to 4 years), rhesus monkeys show no loss of ability to solve these particular problems such as the losses of learning abilities described by Scott (1962) in his analysis of the critical-period concept.

These data do not preclude the possibility of finding other types of learning or learning situations that might be influenced by critical periods of development. The existing categories of learning tasks described in this survey may be inadequate for describing the course of development of many other learning abilities. For example, a monkey that may require lengthy pretraining and adaptation to an apparatus as well as 20 to 100 trials to solve one two-choice object-discrimination problem will, in a matter of seconds or, at most, minutes, become thoroughly adapted to a particular dominance status when introduced for the first time into a social situation with three or four cagemates. At the other extreme, we might find monkeys capable of solving the most complex laboratory problem, but incapable of learning how to respond appropriately to a cagemate of either sex. Studies of social development (Mason, 1961b) have suggested that learning is probably involved in the organization of monkey social behavior, but that there may be a critical period of development during which this learning must take place. Thus, any decision concerning critical periods in the development of learning in the infant monkey must be left tentative until learning situations other than the traditional laboratory tasks are analyzed.

The neonatal and infant rhesus monkey has been tested in almost all of the standard learning situations devised by comparative psychologists. The scientists associated with this research have been extremely fortunate to find an organism with the ability to adapt to these "man-made" methods of studying the learning process. In recognition of these efforts we would like to nominate the infant rhesus monkey as the "standardized infant primate" for future interspecies comparisons, in spite of the fact that a reasonably well-developed theory of behavior has not been worked out for this primate voted "most likely to succeed."

REFERENCES

Birch, H. G. (1945). The role of motivational factors in insightful problem-solving. *J. comp. Psychol.* **38**, 295.

Bitterman, M. E. (1960). Toward a comparative psychology of learning. *Amer. Psychologist* **15**, 704.

Bitterman, M. E. (1962). Techniques for the study of learning in animals: analysis and classification. *Psychol. Bull.* **59**, 81.

Blomquist, A. J., & Harlow, H. F. (1961). The infant rhesus monkey program at the University of Wisconsin Primate Laboratory. *Proc. Anim. Care Panel* **11**, 57.

Carpenter, C. R. (1934). A field study of the behavior and social relations of howling monkeys (*Alouatta palliata*). *Comp. Psychol. Monogr.* **10**, No. 2 (Whole No. 48).

Carpenter, C. R. (1935). Behavior of red spider monkeys in Panama. *J. Mammal.* **16,** 171.

Carpenter, C. R. (1940). A field study in Siam of the behavior and social relations of the gibbon (*Hylobates lar*). *Comp. Psychol. Monogr.* **16,** No. 5 (Whole No. 84).

Davis, R. T., McDowell, A. A., & Nissen, H. W. (1957). Solution of bent-wire problems by monkeys and chimpanzees. *J. comp. physiol. Psychol.* **50,** 441.

Finch, G. (1941). The solution of patterned string problems by chimpanzees. *J. comp. Psychol.* **32,** 83.

Fitzgerald, Alice (1935). Rearing marmosets in captivity. *J. Mammal.* **16,** 181.

Foley, J. P., Jr. (1934). First year development of a rhesus monkey (*Macaca mulatta*) reared in isolation. *J. genet. Psychol.* **45,** 39.

Green, P. C. (1962). Learning, extinction, and generalization of conditioned responses by young monkeys. *Psychol. Rep.* **10,** 731.

Grether, W. F. (1939). Color vision and color blindness in monkeys. *Comp. Psychol. Monogr.* **15,** No. 4 (Whole No. 76).

Harlow, H. F. (1951). Primate learning. *In* "Comparative Psychology," (C. P. Stone, ed.), 3rd ed., pp. 183–238. Prentice-Hall, New York.

Harlow, H. F. (1958). The evolution of learning. *In* "Behavior and Evolution" (Anne Roe & G. G. Simpson, eds.), pp. 269–290. Yale Univer. Press, New Haven, Connecticut.

Harlow, H. F. (1959). The development of learning in the rhesus monkey. *Amer. Scientist* **47,** 458. (Reprinted in "Science in Progress" [W. R. Brode, ed.], 12th series, pp. 239–269. Yale Univer. Press, New Haven, Connecticut, 1962.)

Harlow, H. F., Blazek, Nancy C., & McClearn, G. E. (1956). Manipulatory motivation in the infant rhesus monkey. *J. comp. physiol. Psychol.* **49,** 444.

Harlow, H. F., Harlow, Margaret K., Rueping, R. R., & Mason, W. A. (1960). Performance of infant rhesus monkeys on discrimination learning, delayed response, and discrimination learning set. *J. comp. physiol. Psychol.* **53,** 113.

Hayes, K. J., & Hayes, Catherine (1952). Imitation in a home-raised chimpanzee. *J. comp. physiol. Psychol.* **45,** 450.

Hayes, K. J., Thompson, R., & Hayes, Catherine (1953). Discrimination learning set in chimpanzees. *J. comp. physiol. Psychol.* **46,** 99.

Hill, W. C. O. (1960). "Primates: Comparative Anatomy and Taxonomy," Vol. 4. Interscience, New York.

Hill, W. C. O. (1962). "Primates: Comparative Anatomy and Taxonomy," Vol. 5. Interscience, New York.

Hines, Marion (1942). The development and regression of reflexes, postures, and progression in the young macaque. *Contrib. Embryol., Carnegie Inst.* **30,** 153.

Jacobsen, C. F., Jacobsen, Marion M., & Yoshioka, J. G. (1932). Development of an infant chimpanzee during her first year. *Comp. Psychol. Monogr.* **9,** No. 1 (Whole No. 41).

Jensen, G. D. (1961). The development of prehension in a macaque. *J. comp. physiol. Psychol.* **54,** 11.

Jolly, Alison (1964). Choice of cue in prosimian learning. *Anim. Behav.* **12,** 571.

Kellogg, W. N., & Kellogg, Laverne A. (1933). "The Ape and the Child." McGraw-Hill, New York.

Kendler, H. H., & Kendler, Tracy S. (1962). Vertical and horizontal processes in problem solving. *Psychol. Rev.* **69,** 1.

Knobloch, Hilda, & Pasamanick, B. (1959a). Gross motor behavior in an infant gorilla. *J. comp. physiol. Psychol.* **52,** 559.

Knobloch, Hilda, & Pasamanick, B. (1959b). The development of adaptive behavior in an infant gorilla. *J. comp. physiol. Psychol.* 52, 699.

Köhler, W. (1925). "The Mentality of Apes." Harcourt, Brace, New York. (Reprinted by Humanities Press, New York, 1951.)

Lucas, N. S., Hume, E. M., & Smith, H. H. (1927). On the breeding of the common marmoset *Hapale jacchus* Linn. in captivity when irradiated with ultraviolet rays. *Proc. zool. Soc. Lond.* p. 447.

Mason, W. A. (1961a). Effects of age and stimulus characteristics on manipulatory responsiveness of monkeys raised in a restricted environment. *J. genet. Psychol.* 99, 301.

Mason, W. A. (1961b). The effects of social restriction on the behavior of rhesus monkeys: III. Dominance tests. *J. comp. physiol. Psychol.* 54, 694.

Mason, W. A., & Harlow, H. F. (1958a). Formation of conditioned responses in infant monkeys. *J. comp. physiol. Psychol.* 51, 68.

Mason, W. A., & Harlow, H. F. (1958b). Performance of infant rhesus monkeys on a spatial discrimination problem. *J. comp. physiol. Psychol.* 51, 71.

Mason, W. A., & Harlow, H. F. (1958c). Learned approach by infant rhesus monkeys to the sucking situation. *Psychol. Rep.* 4, 79.

Mason, W. A., & Harlow, H. F. (1959). Initial responses of infant rhesus monkeys to solid foods. *Psychol. Rep.* 5, 193.

Mason, W. A., & Harlow, H. F. (1961). The effects of age and previous training on patterned-strings performance of rhesus monkeys. *J. comp. physiol. Psychol.* 54, 704.

Mason, W. A., Blazek, Nancy C., & Harlow, H. F. (1956). Learning capacities of the infant rhesus monkey. *J. comp. physiol. Psychol.* 49, 449.

Nolte, Angela F. (1955). Observations on the behavior of free ranging *Macaca radiata* in southern India. *Z. Tierpsychol.* 12, 77.

Reese, H. W. (1963). Discrimination learning set in children. *Advanc. child Develpm. Behav.* 1, 115.

Richter, C. P. (1931). The grasping reflex in the new-born monkey. *Arch. Neurol. Psychiat.* 26, 784.

Riesen, A. H., & Kinder, Elaine F. (1952). "Postural Development of Infant Chimpanzees." Yale Univer. Press, New Haven, Connecticut.

Riesen, A. H., Greenberg, B., Granston, A. S., & Fantz, R. L. (1953). Solutions of patterned string problems by young gorillas. *J. comp. physiol. Psychol.* 46, 19.

Rumbaugh, D. M., & Rice, Carol P. (1962). Learning-set formation in young great apes. *J. comp. physiol. Psychol.* 55, 866.

Schiller, P. H. (1952). Innate constituents of complex responses in primates. *Psychol. Rev.* 59, 177.

Scott, J. P. (1962). Critical periods in behavioral development. *Science* 138, 949.

Solenkova, E. G., & Nikitina, G. M. (1960). The initial formation and development of conditioned defensive reflexes in the young of monkeys. *Pavlov J. higher nerv. Activ.* 10, 220.

Tighe, T. J. (1964). Reversal and nonreversal shifts in monkeys. *J. comp. physiol. Psychol.* 58, 324.

Torrey, Natalie E. (1963). Learning set formation in the infant monkey (*Macaca mulatta*). Doctoral dissertation, Cornell University. University Microfilms, Ann Arbor, Michigan, No. 64-3691.

Warden, C. J. (1927). "A Short Outline of Comparative Psychology." Norton, New York.

Yerkes, R. M., & Tomilin, M. I. (1935). Mother-infant relations in chimpanzee. *J. comp. Psychol.* **20**, 321.

Zimmermann, R. R. (1958). Analysis of discrimination learning in infant rhesus monkeys. Doctoral dissertation, University of Wisconsin. University Microfilms, Ann Arbor, Michigan, No. 58–7554.

Zimmermann, R. R. (1961). Analysis of discrimination learning capacities in the infant rhesus monkey. *J. comp. physiol. Psychol.* **54**, 1.

Zimmermann, R. R. (1962). Form generalization in the infant monkey. *J. comp. physiol. Psychol.* **55**, 918.

Zimmermann, R. R. (1963). Brightness transposition in the infant and adult rhesus monkey. *Amer. Psychologist* **18**, 472. (Abstract.)

Chapter 12

Age Changes in Chimpanzees[1]

A. J. Riopelle and C. M. Rogers

Delta Regional Primate Research Center, Covington, Louisiana, and Yerkes Laboratories of Primate Biology, Orange Park, Florida

I. INTRODUCTION

It is amazing that no laboratory has studied primates throughout their lifetimes, despite the large number of laboratories, particularly in the United States, in which primates are used for research. Studies of how behavior changes with age are regrettably unfashionable, probably because behavioral investigators abhor biological factors. We frequently encountered the opinion that studying old chimpanzees (*Pan*) is like running a hotel for ancient apes. But Henry Nissen long ago realized the importance of lifetime studies and was determined to maintain old animals despite their unpopularity. Thanks to his foresight, we are able to report the data presented in this chapter.

Fortunately, there is developing now a broad awareness of the need for more intimate study of the higher primates throughout their entire life spans, together with a recognition that lower animals often are inadequate experimental subjects for the study of human problems of

[1]Supported in part by grants (HE-05383 and H-5691) from the National Heart Institute, U. S. Public Health Service, and in part by Contract No. AT-(40-1)-1553 with the U. S. Atomic Energy Commission.

health and behavior. Doubtless the rotifer has many advantages for studying aging, most of which relate to rapidity of reproduction, convenience of maintenance, or the ease with which it can be sliced for studies under the electron microscope. But it would be bold indeed to extrapolate from an animal with a life span of 42 days to one like the human having a life span of 100 years. The white rat, too, has its virtues, among them a life span within the present socially acceptable period for a grant from a recognized agency. But the behavior of a naive albino rat is truly limited in comparison with the behavior of a primate. Reliance on this species in psychology laboratories has resulted in thorough investigations of situations important for humans such as starvation to 80% body weight, 23-hour water deprivation, immersion in water to over-head depths and electric shock to the point where electrocution is a consideration.

Because of the many similarities of chimpanzees to man—social, psychological, endocrinological, and anatomical—this animal should be especially useful for studying problems of great human significance. Aging is one such problem.

Our interest in using chimpanzees for studies of aging stemmed from a number of sources. First, the Yerkes Laboratories have been committed to a study of the psychobiological development of the chimpanzee throughout its lifetime, and maturity and senescence had yet to be studied. Second, we already had in the Laboratories some animals, the "ancient apes" to which many have referred, and it became possible to make such studies. Third, we wanted to know if the kinds of behavioral changes that occur with old age mimicked changes induced by experimental insults like removal of various parts of the brain. Finally, the Laboratories have been engaged over the past 8 years in a study on the effects of ionizing radiation (see Chapter 14 by Davis). In view of presumed relations between irradiation and aging we believed it would be important to get some idea of the changes wrought by age in order to anticipate the alterations due to irradiation.

II. LIFE HISTORY

A. Infancy and Prepuberty

The infant chimpanzee is born after a gestation period that averages 226 days. At birth, the average weight of the chimpanzee is about 1.8 kg or 3.5 lb. All deciduous teeth have erupted by the age of about 12 months in the chimpanzee, compared to about 28 months in the human. The period of helplessness and complete dependence lasts about 2 years in the laboratory chimpanzee compared with 4 or 5 years in the human.

The period from then until puberty may be characterized by relatively carefree activity, boundless energy, rapid shifts of mood, and great sociability toward humans. Juvenile chimpanzees may be picked up, carried about, and handled with relative impunity, particularly if they are in daily contact with humans. Because of the ease with which juveniles can be handled, they are the most widely sought subjects for participation in experiments (see Chapter 9 by Mason).

B. Adolescence

A significant period in the life of every animal is the transition to sexual maturity. This transition is marked by important primary hormonal changes which lead to a number of secondary changes, some of which are subtle and others quite apparent. The development of pubic and axillary hair is well known in the human as is also the enlargement of the mammae in females and the voice change in males. The most easily defined point in pubertal development is the onset of menstruation which, in humans, occurs at about 13.5 years. The chimpanzee reaches the same state of maturity in an average of 8.8 years, or in roughly two-thirds of the time required by the human. The occurrence of the male counterpart, the production of sperm, is not known with precision in either the human or the chimpanzee. In humans, however, it lags behind menstruation by months or years. If changes in hair pattern attributable to androgen production occur in chimpanzees, they are unfortunately so subtle that they have not been characterized. Also, there are no dramatic alterations in voice character in chimpanzees. Because of the absence of clear changes, it is difficult to establish puberty in the male chimpanzee. A growth spurt does exist in chimpanzees as it does in humans. It begins at age 7 and continues until age 11 or 12.

During adolescence, several other expressions of maturation occur which may be used to compare the human with the chimpanzee. The ages at which molars erupt and the epiphyses of the long bones close typically are 55% to 70% as great for the chimpanzee as for the human. These measures represent transition points within adolescence and serve to translate an age scale into a scale of maturity for the two species.

A number of behavioral changes occur during adolescence. The spontaneous and nearly continuous playful activity of the juvenile decreases and separates into discrete units; hence there appear to be more acts and less activity. Acts are interspersed among periods of quiet and inactivity. Such periods may be pierced by vocal outbursts, and feeding elicits raucous noise (see Chapter 15 by Jay). Prominent also are marked sex differences in the character and vigor of action. When startled or greeting an acquaintance or stranger, a male chimpanzee claps his hands, stamps his feet on the nearest resonant metal plate, or dances around his cage, warming to a frenzied outburst that may cease even more abruptly than it began. Other male animals drum on the walls, swing from place to place, or even spit water or throw feces at the intruder. Females, on the other hand, may remain quiet or approach the cage wire directly to await an opportunity for a surprise attack. These differential behavioral characteristics persist through most, if not all, of adult life.

C. Old Age

The transition from vigorous adulthood to old age requires many years and there are few abrupt events to serve as demarcation points. One which beckons for attention is the termination of the reproductive period—the beginning of the menopause. This phenomenon occurs in women between 35 and 50 years of age, long before senility is evident. Surely, few women still capable of reproduction would be considered biologically old. The female chimpanzees described as "old" in this chapter, however, are still menstruating, and one conceived at age 40. This single fact disagrees with all other evidence we have obtained which can serve as an index of maturation. However, Asdell (1946) concluded that man is the only species studied thus far which has a true menopause. Therefore, Riopelle (1963) has rejected the notion that animals which still bear young are themselves young.

X-ray photographs and microscopic structure of bones provide more direct evidence that a 40-year-old chimpanzee is comparable to a human of 60 to 80 years. Riopelle (1963) extrapolated from these observations plus eruption of permanent teeth and closure of epiphyses, and concluded that the maximum life span of the chimpanzee is about 60 years. This would correspond to a human age of 90 to 100 years. This estimate, incidentally, is substantially greater than those put forth previously.

Old age in chimpanzees, as we have observed it in five females and one male over 30 years of age, is characterized by a further reduction in activity, reduced vigor of motion, slower locomotion, more bipedal walking, and more self-grooming.

The oldest chimpanzee in the Yerkes Laboratories is a female named Pati, believed to be more than 40 years old. She came as a gift from Señora Rosalia Abreu, of Havana, Cuba, who obtained her at the approximate age of 5 (when ages are easy to determine with precision). She is undoubtedy one of the oldest chimpanzees in the world and now displays the clearest signs of old age. Her ambulatory movements are usually bipedal, she avoids social contact with humans, and she spends a great deal of her time in activity directed toward herself or in solitary activity. Self-grooming, which she does by parting the hair with her teeth and pulling out loose hairs with her lips, occupies a disproportionately great amount of time. Like all of the older animals, during the winter she usually stays indoors where it is heated.

Her cage partner, Wendy, age 39, shows less upright walking and self-grooming but grooms Pati frequently. Her movements are quicker than Pati's, and the staff generally holds that she could still defend herself admirably. Soda and Vera, although aged 35 and 30 respectively, seem to be between Pati and Wendy biologically. They, too, show deliberateness of action and disdain for humans.

Unfortunately, all chimpanzees in the Laboratories over the age of 30 are females. A healthy male of 35 years died in 1960 when he was restrained on an animal board during some experiments in which blood was withdrawn and pyrogenic agents were injected. This animal, named Bokar, had a useful if not long life. He had sired 40 offspring and had participated in many researches. Shortly after his demise, one of the staff members commented regretfully about poor Bokar's fate. He was admonished by another colleague not to feel too bad; after all, Bokar had more publications than anyone else in the Laboratories!

III. PERFORMANCE ON FORMAL TESTS OF BEHAVIOR

Gross activity of chimpanzees gradually declines from youth to maturity to old age. We turn now to performance on formal tests of behavioral capacity to see if a similar trend can be discerned. We sought to determine whether or not performance changed systematically with age and whether or not a single factor could account for the changes. The tests selected, although diverse in nature, were fairly similar in operation. The chimpanzee had merely to displace a stimulus object from a tray to obtain food reward (Fig. 1). The apparatus, the same type as that used at the Yerkes Laboratories for over 25 years, was wheeled to each animal's cage. All baiting took place behind an opaque screen to prevent inadvertent cues.

The 19 chimpanzees used as subjects ranged in age from 7 to 41 years, which covers the period from prepuberty to old age. Some were wild-born, others born in captivity. The oldest captive-born animals were Bula and Alpha, both age 30. Alpha was the first chimpanzee born at the Yerkes Laboratories and Bula was born at the Rosalia Abreu Estate in Havana, Cuba.

A. Consecutive Discriminations

In the first test, 200 two-choice simultaneous-discrimination problems were given at a rate of 16 per day. Stimuli were of the diverse varieties often used in studies of learning-set formation (see Chapter 2 by Miles in Volume I). Each problem lasted for four trials, after which the stimuli

Fig. 1. Adult chimpanzee displacing stimulus object from foodwell on a retractable tray. (From Riopelle, 1962.)

were changed. The data were summarized in terms of the percentage of correct responses on the second trials of problems 151 to 200, since asymptotic performance was assumed to be least affected by differences in prior experience.

The results are presented in Fig. 2, which shows the performance of each chimpanzee and the least-squares regression line of performance on age. In contrast to the original expectation of a decline following puberty or early maturity, there was a consistent though nonsignificant improvement in performance with age. No particular theoretical importance should be given to the form or parameters of the equation of the line. It merely represents a reasonable estimate of the trend of the data.

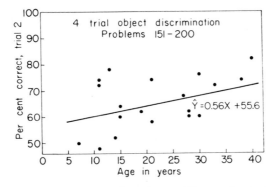

FIG. 2. Improvement with age of performance on 50 easy discrimination problems presented consecutively. (After Riopelle & Rogers, 1963.)

B. Concurrent Discriminations

In the second test, 10 new discrimination problems were presented each day. On each trial, the chimpanzees were required to select one of two objects, as before. There was one important difference, however. In the earlier test the four trials of each problem were presented consecutively, whereas in the present test all 10 of the day's pairs of objects were presented once in serial order; then the sequence was repeated nine more times. Thus, nine problems were always interposed between consecutive presentations of any given problem. Obviously, this adds greatly to the interproblem interference. The older animals' performances again equaled those of the younger chimpanzees (Fig. 3).

C. Difficult Pattern Discriminations

The discrimination problems described above placed few demands on visual discriminative capacity. The stimuli were many times larger than the minimum visible, they differed in size, and they differed in hue and shape. The next tests were selected because abundant evidence (see Chapter 1 by Meyer *et al.* in Volume I) indicated that they were much more difficult than the problems described above. Discriminations were of the following types: Dissimilar Forms (two different forms cut from plywood and painted the same color), Dissimilar Patterns (two different patterns painted on white wedges), Mirror-Image Forms, and Mirror-Image Patterns. The usual differences among tests were clearly evident. On none of the tests, however, could one discern a trend toward improvement or decline with age.

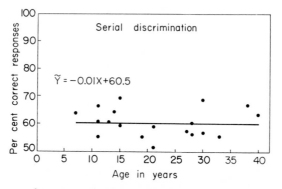

Fig. 3. Percentage of correct responses on runs 2 to 10 of visual discrimination problems presented concurrently. (After Riopelle & Rogers, 1963.)

D. Spatial Delayed Response

The next test administered was a spatial delayed-response test using the direct method of baiting (see Chapter 4 by Fletcher in Volume I). Five identical black squares were presented to the animal on a horizontal tray. Each of the squares covered a foodwell. One of the five, selected at random, was raised and the foodwell beneath it baited in full view of the subject. Then, after a brief delay, the tray bearing the squares was pushed toward the chimpanzee and it was permitted to respond. Thirty-three trials were given daily.

Figure 4 shows the percentage of correct responses after an imposed 5-second delay. The older chimpanzees performed more poorly on this test than the younger chimpanzees. The same trend resulted when the tray was pushed forward as soon as possible after the foodwell was baited (0-second delay), indicating that the decrement may be one of attention rather than retention. Increasing the delay to 10 seconds depressed the performance of young animals more than that of the old animals. Performance was not worse than that of the older animals, however; the downward slope was merely less steep. This fact suggests that the younger chimpanzees become distracted more readily during the longer interval, an expression, no doubt, of the juveniles' greater activity, alertness to, and interest in events of the surrounding environment.

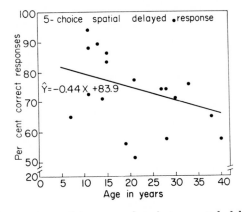

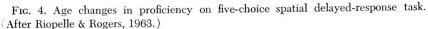

FIG. 4. Age changes in proficiency on five-choice spatial delayed-response task. (After Riopelle & Rogers, 1963.)

E. Four-Choice Oddity

In the next test, four vertical panels were presented on each trial. Three of the panels were identical and the fourth different. Food was always located behind the odd panel and could be obtained by simply pushing back the suspended pattern.

Again there was a clear decline in performance with age (Fig. 5), but the decline followed a course radically different from that for delayed response. Between the ages of 7 and 20, performance declines rapidly;

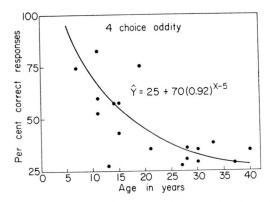

FIG. 5. Decrement in performance on four-choice oddity discriminations.

after that no change occurs. The data can be described by an exponential curve in which performance drops about 8% of the remaining distance per year. It should be emphasized again that these are empirical, not theoretical, trends, and no attempt should be made to extrapolate the curve toward zero age, for to do so would predict performance better than 100% correct.

F. Novelty and Perseveration

Adult humans have long been thought to become more rigid and inflexible with the passing of years. Much objective evidence supporting this statement can be found in studies by Ruch (1933), Schaie (1958), and others. Other data show political and social conservatism. In our investigations, which bear only remotely on this topic, we tested the hypothesis that old chimpanzees would tend to select a familiar stimulus object rather than an unfamiliar one. We presented 10 pairs of objects to the animals each day for 10 days. The chimpanzees had seen one of the objects of each pair on many previous occasions, but the other object was totally novel. The familiar and the unfamiliar object covered food equally often. Thus the selection of one or the other led to equivalent outcomes. The question is: Did the old chimpanzees tend to select the familiar stimulus and the younger animals the unfamiliar stimulus? On the basis of the excellent work of Berlyne (1950) on exploratory behavior, one would anticipate a tendency for all animals to select the novel stimulus. On the other hand, juvenile rhesus monkeys (*Macaca mulatta*) preferred the familiar stimulus for a considerable period during a similar experiment (Riopelle *et al.*, 1962).

The data for chimpanzees are presented in Fig. 6 in terms of the percentage of trials on which the familiar stimulus was chosen. Percentages greater than 50 indicate a tendency to choose the familiar stimulus, whereas any percentages below 50 indicate selection of the unfamiliar stimulus. As can be seen from this figure, preference was unrelated to

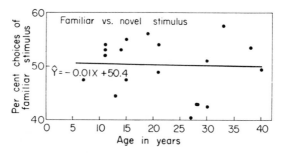

Fig. 6. Percentage of trials on which a familiar stimulus was chosen when paired with a novel stimulus.

age and individual differences were small. Preferences were very weak; no animal selected the familiar stimulus on fewer than 40% nor more than 60% of the trials.

Corroborative evidence comes from a related experiment in which a stimulus to which responses were consistently rewarded was paired with one to which responses were variably rewarded. As above, two objects were presented, both for 100 trials. One of the stimulus objects invariably covered two pieces of grape; the other object covered 0, 1, 2, 3, or 4 pieces of grape on an unpredictable schedule. Although the over-all rewards were identical, we thought that animals of different ages might adopt different strategies. Such was not the case, as can be seen in Fig. 7.

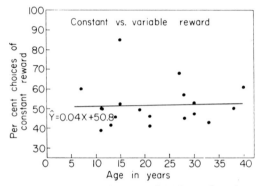

Fig. 7. Percentage of trials on which a stimulus object denoting a constant reward was preferred to another stimulus object which yielded a variable reward.

Performances were entirely unrelated to age. One chimpanzee strongly preferred the constantly rewarded stimulus, but his age was only 15 years.

Studies related to these were recently conducted by Bernstein (1961), who was interested in variability of response. He used 16 chimpanzees, divided into two age groups: 11 to 19 and 28 to 40 years. In one experiment, three objects were presented to the chimpanzees. Two of the objects always covered baited foodwells, while the third always covered an empty foodwell. Only a single object-displacement was permitted on any trial. Bernstein measured the consistency of response to the different objects. He found that most animals, regardless of age, alternated frequently between the two objects covering food.

In a second experiment, Bernstein measured response variability when increased effort was the penalty for a perseverative error. Food was placed in various positions on a horizontal tray which could be rotated to bring the food within reach. One exceptional animal always turned the wheel counterclockwise regardless of the position of the food whenever it could not be obtained by direct reaching. But most chimpanzees brought the food by the shortest route except when the food was in a sector on the far side of the tray, in which case the direction of response was variable. The width of the sector producing variable responding did not differ for the two age groups.

Bernstein also found no age differences in conditional discrimination or in discrimination reversal. Flexibility of response is of paramount importance for proficient performance on both tasks, since both require the subject to cease selecting one stimulus and to begin selecting another (see Chapter 5 by French in Volume I). These results together with those described above suggest that older chimpanzees do not respond more rigidly than do younger ones.

IV. DISCUSSION

There are many similarities between the chimpanzee and the human. Both have a long infancy, an exuberant childhood, and a decline in overt activity accompanied by an increase in sex differences during maturity. As old age is reached, the chimpanzee, like the human, becomes more interested in itself and initiates fewer externally-directed activities. When examined on formal behavioral tests, some chimpanzee performances improve with age, whereas others fall. Declines, when found, manifest themselves in early maturity, as is also true in humans (Jones, 1959). For this reason, we need not specify with precision the onset or duration of "old age." When many graphs of performance are assembled in one place, one cannot help being impressed by how early in life the decline of performance actually begins, especially if one has already passed that age!

One of the difficulties of using human subjects for measuring changes in capacity is the possible confounding of motivational changes with changes in capacity. Decrements in human performance with age may often be attributed to changed motivation rather than lessened ability. The losses of chimpanzees on the delayed-response and oddity tasks are not likely to be explained by this factor, however, since the response was not fatiguing and the pace was not forced, and decrements did not occur on other tasks requiring the same kind of response; that is, displacement of stimulus objects. Also, the chimpanzees were not tested under food deprivation, nor were they forced to be tested. Any animal could have walked away at any time. However, the chimpanzees appreciated the social contact of testing and we had visible signs of high motivation in their excitement, pleasure panting, and even masturbation as the test apparatus was wheeled toward their cages. Typically, the chimpanzee was positioned for testing before the apparatus arrived.

The improvement with age in performance on consecutive discriminations might be attributable to the cumulative effects of prior test experience. However, there is little reason to believe that this same cumulative experience would not improve performance on the delayed-response and oddity tests. But attributing the losses on these two tests to waning motivation or loss of discrimination would leave the embarrassing problem of the preservation of skill on the other tasks.

The tests employed here have also been useful for analyzing behavior after excision of cortical tissue in primates, and the results are pertinent for the present findings. A characteristic result of those studies is a loss on delayed response following removal of portions of the prefrontal lobes (see Chapter 4 by Fletcher in Volume I). Preservation of the visual discriminative functions is, in contrast, also characteristic (Rosvold et al., 1961). On the other hand, removal of the temporal lobes of monkeys impairs discrimination performance but not proficiency on delayed response (Pribram & Bagshaw, 1953). The oddity test is sensitive to lesions in both the frontal and the posterior association areas (Harlow et al., 1951). The patterns of loss in the present findings suggest, therefore, a decline with age in the functional integrity of the prefrontal-lobe system.

One further finding may be noted. Several animals of our group had been subjected to 375 r of X- or γ-radiation. For all the tests described in this chapter, their performances were compared with those of the other animals. In no case could we find evidence that the irradiated animals differed from the nonirradiated. Their performances, instead, were in accordance with their ages. Accelerated aging has been thought due to radiation pathology. The present data give little comfort to this conclusion (see also Chapter 14 by Davis).

Our analysis of the performance characteristics of aged chimpanzees is scarcely begun; however, the results obtained thus far are encouraging. It is already clear that performance changes are quite specific and differ markedly from task to task. No single equation describes all the data.

References

Asdell, S. A. (1946). "Patterns of Mammalian Reproduction." Comstock, Ithaca, New York.

Berlyne, D. E. (1950). Novelty and curiosity as determinants of exploratory behavior. Brit. J. Psychol. **41**, 68.

Bernstein, I. S. (1961). Response variability and rigidity in the adult chimpanzee. J. Geront. **16**, 381.

Harlow, H. F., Meyer, D. R., & Settlage, P. H. (1951). The effects of large cortical lesions on the solution of oddity problems by monkeys. J. comp. physiol. Psychol. **44**, 320.

Jones, H. E. (1959). Intelligence and problem solving. *In* "Handbook of Aging and the Individual" (J. E. Birren, ed.), pp. 700–738. Univer. Chicago Press, Chicago, Illinois.

Pribram, K. H., & Bagshaw, Muriel (1953). Further analysis of the temporal lobe syndrom utilizing frontotemporal ablations. *J. comp. Neurol.* 99, 347.

Riopelle, A. J. (1962). Some behavioral effects of ionizing radiation on primates. *In* "Response of the Nervous System to Ionizing Radiation" (T. J. Haley & R. S. Snider, eds.), pp. 719–728. Academic Press, New York.

Riopelle, A. J. (1963). Growth and behavioral changes in chimpanzees. *Z. Morphol. Anthropol.* 53, 53.

Riopelle, A. J., & Rogers, C. M. (1963). Behavior of chimpanzees of differing ages. *Activ. nerv. superior* 5, 260.

Riopelle, A. J., Cronholm, J. N., & Addison, R. G. (1962). Stimulus familiarity and multiple discrimination learning. *J. comp. physiol. Psychol.* 55, 274.

Rosvold, H. E., Szwarcbart, Maria K., Mirsky, A. F., & Mishkin, M. (1961). The effect of frontal-lobe damage on delayed-response performance in chimpanzees. *J. comp. physiol. Psychol.* 54, 368.

Ruch, F. L. (1933). Adult learning. *Psychol. Bull.* 30, 387.

Schaie, K. W. (1958). Rigidity-flexibility and intelligence: a cross-sectional study of the adult life span from 20 to 70 years. *Psychol. Monogr.* 72, No. 9 (Whole No. 462).

Chapter 13

Investigative Behavior

Robert A. Butler

Departments of Surgery and Psychology, University of Chicago, Chicago, Illinois

I. INTRODUCTION

A. Pre-experimental Observations

"Mr. Small notes the inveterate tendency of his rats to 'fool.' They run about, smell, dig, or gnaw, without real reference to the business in hand" (Hobhouse, 1901, p. 195). "The business in hand" happened to be finding a hidden entrance to a box containing food. In his experiments on problem solving, Hobhouse, too, noted those "native" actions of several different animals. He spoke, for example, of the kitten wandering about and picking at things, the elephant ceaselessly fumbling, and the monkey pulling things about. Hobhouse thought that these behaviors became of use through experience; yet, irrespective of the amount of experience, when "intelligence fails" an animal might revert to such aimless activities. These were, he concluded, a hindrance to the work of intelligence. Small (1900), on the other hand, was most impressed by what others considered purposeless behavior of his rats. He unreservedly maintained that this "tendency to fool" was an expression of curiosity, contending that it was a "fundamental and irreducible desire to know all this new environment" (Small, 1900, p. 214). Small is remembered today as the man who first used the maze to study animal learning, but he was also deeply interested in the motivational determinants of animal behavior. He suggested that the curiosity which impels an animal to become familiar with its locale becomes, in man, the basis for artistic and scientific achievements. He urged that a search be made for the "relatively invariable factors in the animal mind," which he believed would be found through "the study of the instinctive traits and tendencies, out of which, in higher differentiations, human nature is made" (Small, 1900, p. 222).

Immediately following Darwin's momentous work, the idea of invariant factors of the animal mind, or mental continuity, became of prime importance to students of natural history. The mental life of nonhuman primates was singled out for special attention. Implicit, at times explicit, in these earlier writings was the notion that curiosity was an intellectual trait developed to a greater extent in monkeys and apes than in lower animals. The famous comparative psychologist, George J. Romanes, observed for several days a hooded capuchin (*Cebus apella*). He described in detail the monkey's behavior toward sundry objects including a lock and key and a brush which could be disassembled and reassembled. Romanes (1892, pp. 483–498) emphasized the fact that the monkey would work diligently on a task which provided no material benefit; the animal seemed to be bent on simply finishing a task. In summarizing his observations, Romanes wrote that "the most striking feature in the psychology of this animal, and one which is least like anything met in other animals, was the tireless spirit of investigation" (Romanes, 1892, p. 487). Dennis (1955) has redirected attention to the passage just quoted and also to an observation by Thorndike (1901) on the behavior of one of his cebus monkeys: "No. 1 happened to hit a projecting wire so as to make it vibrate. He repeated this act hundreds of times in the few days following. He did not, could not, eat, make love to or get preliminary practice for the serious battles of life out of that sound. But it did give him mental food, mental exercise" (Thorndike, 1901, p. 55). The monkey "likes to be active for the sake of activity" (Thorndike, 1901, p. 55).

Other investigators, of course, commented on the inquisitive nature of animals, especially that of monkeys and apes. Revesz (1925), for one, was convinced that the monkey's interest extended "far beyond the satisfaction of elementary needs" (Revesz, 1925, p. 314). From his observations on the manipulative behavior of two rhesus monkeys (*Macaca mulatta*) and a mangabey (*Cercocebus torquatus*), he concluded that these animals had an interest in objects as such, in their constituents and functions. Yerkes and Petrunkevitch (1925) pointed out that Mrs. Kohts, in her pioneer work on the visual capacities of the chimpanzee (*Pan*), discovered that the opportunity to play served effectively as a reward for satisfactory performance on visual discrimination problems. The chimpanzee's playfulness and its omnipotent curiosity is mentioned repeatedly in her book, *Infant Ape and Human Child* (Kohts, 1935). Here Mrs. Kohts compares her notes on the chimpanzee's behavior with those made later on the activities of her child. Both the child and the chimpanzee liked to grasp brightly colored objects. They almost always worked at things that could be moved easily, such as doors, windows, rotating stools, piano lids, and the like. They also intently watched things that moved. A visit to the city would provoke eager attention to the great variety of events. Kellogg and Kellogg (1933) reared their child and an infant chimpanzee together, thus affording a more precise comparison of the responsiveness of the two infants to the environment. With regard to curiosity, the Kelloggs' remarks are similar to those of Mrs. Kohts. Notably, the child and the chimpanzee were highly alike in their attentiveness to new objects, toys, and pictures. Both derived obvious enjoyment from play and companionship. It hardly seems necessary to point out the differences between an infant ape and a child since they are, indeed, enormous. Within the sphere of manipulative and curiosity behavior, Mrs. Kohts emphasized the greater competence of the child in use of tools and his predilection for constructive play. The Kelloggs pointed out the child's greater attention span and his appreciation of the larger aspects of the environment—the over-all situation.

The rhesus and cebus monkeys and the chimpanzee have been studied far more extensively than any of the other nonhuman primates. While their investigative activities have been mentioned in various papers, the reports were usually brief and incidental to the expressed purposes of the researchers. Still, the descriptions of investigative behavior are much more abundant for these than for other nonhuman primates. Occasionally, however, curiosity behavior in the less-studied primates has been mentioned. Lowther (1940), after observing two galagos (*Galago senegalensis*) for several years, contended that they are ". . . as inquisitive as a monkey. Unlike that animal, however, curiosity manifests itself by smelling instead of handling the strange object" (Lowther, 1940, p. 446). The marmoset (*Callithrix jacchus*), according to Fitzgerald (1935), is a rabid visual explorer and ". . . any change in the position of an object, or any addition to the furnishings of the room is at once investigated" (Fitzgerald, 1935, p. 182). Several field studies of monkeys and apes include an incidental reference to the curiosity of the species being observed.

The consensus is that primates are an inquisitive lot. Indeed, this opinion has become so pervasive that a primate which appears indifferent to the procession of events is considered an anomaly. For example, Yerkes (1927), in his study of the gorilla (*Gorilla gorilla*) Congo, reported that ". . . she lacked, by comparison with other types of ape, interest in mechanical devices, tendency to fool with them, or even to manipulate them with definite intent when they appeared in problem-situations" (Yerkes, 1927, p. 186). He reasoned that "either she inhibited expressions of curiosity or she is decidedly less curious about strange or novel objects and events than are other apes" (Yerkes, 1927, p. 185).

These early workers generally held that the basic curiosity of primates possessed great significance for the evolution of man. Nevertheless, no formal experimentation in this area was carried out. Most animal psychologists were studying learning and they used, almost exclusively, the white rat. In so doing they could not fail to observe its incessant explorations. As a consequence, formal research on investigative behavior was initiated on this animal nearly 25 years before the primate research got underway. Some of the early and important studies on the rat will be discussed briefly to provide a more adequate perspective for the experiments on investigative behavior of primates.

B. Experimentation with Rats

1. Early Studies of Investigative Behavior

Dashiell (1925) described a technique for demonstrating to students the operation of an organic drive. He would place hungry and fed rats in a multiunit maze containing no food. Students, equipped with a floor plan of the maze, would trace the rats' movements during a 60-second period. The hungry group routinely entered more maze units than the group satiated for food. The interpretation of these observations has been that exploration increases the probability of a hungry animal finding food and is, therefore, of survival value. Dashiell remarked, however, that being placed in a new environment will trigger an exploratory reaction whether or not the animal has been fed.

Somewhat later, Nissen (1930) placed rats satiated for food and water in the Columbia obstruction box to find out if they would cross an electrically-charged grid in order to explore. By crossing the grid, the animals could enter a compartmentalized box containing sawdust, wood blocks, etc. He discovered that rats would indeed subject themselves to electric shock in return for a chance to investigate the goal box.

Dennis and Sollenberger (1934) recorded the sequence of choice-point behaviors in simple mazes containing no extrinsic rewards such as food or water. A most prominent feature of the rats' behavior was their choice of that arm of the maze least recently occupied. Dennis and Sollenberger attributed this alternation of choices to an exploratory tendency. Heathers (1940), running rats in a T-maze, also reported alternation in choice-point behavior, but emphasized the response, rather than the stimulus, in accounting for the results. He proposed a response decrement or a tendency to avoid the response most recently performed. At this point, the problem was dropped for a full decade leaving only the general impression that investigative behavior of rats certainly exists and is probably of some survival value. In the confines of the laboratory, it might account for spontaneous alternation in simple mazes, although one could explain this behavior equally well in other ways.

2. REVIVAL OF INTEREST IN EXPLORATION

In the 1950s, experiments on rats, monkeys, and apes were conducted concurrently, with the plan of the later studies influenced by the data from the preceding ones, irrespective of which species was being tested. Before discussing the experimental work on monkeys and apes, three studies with rats will be described. These experiments not only made direct contact with the earlier work on exploration, but served also as prototypes for a large number of later studies. First, Montgomery (1952) investigated whether spontaneous alternation could be more adequately explained in terms of an exploratory tendency or a response decrement. He devised a maze in which the rat would reach the same place by alternating turns. Alternating places, on the other hand, required that the animal make the same turn. Rats tended to alternate places rather than turns, so Montgomery inferred that an exploratory drive was responsible for spontaneous alternation. Another significant study was that of Berlyne (1950), in which one group of rats was exposed to a familiar object and another was exposed to an unfamiliar object. The group confronted by the novel object exhibited more investigative behavior (time spent sniffing the object). Lastly, Girdner (1953) reported that rats will press a lever faster if the response turns on a dim lamp, a phenomenon which has now been studied under a great many conditions.

These studies, although interesting in their own right, probably would not have helped start a new trend in animal research had they not touched upon the adequacy of the notion that drive-reduction is necessary for learning. What drive is reduced when a response is rewarded, say, by a light or a novel scene? Is it necessary to postulate motives other than the so-called primary biological drives to explain these and the comparable behaviors that were being studied in monkeys? Harlow most emphatically answered yes! In his well-known paper, "Mice, Monkeys, Men, and Motives" (Harlow, 1953a), he vigorously contended that behavior can be motivated by exteroceptive as well as by interoceptive stimuli, and that the curiosity-investigative motives play an enormously important role in the minute-to-minute contact with the environment.

Studies on nonhuman primates that stemmed from this line of thinking will now be discussed. They have been influential both in the area of learning theory and in the field of personality development.

II. MAINTENANCE OF BEHAVIOR IN NONHUMAN PRIMATES BY INVESTIGATABLE REWARDS

A. Manipulative Rewards

1. MECHANICAL PUZZLES

Rhesus monkeys will work on simple tasks in which the only apparent reward is the opportunity to manipulate. Harlow *et al.* (1950) were the first to demonstrate this experimentally. The manipulandum in this original situation was a mechanical puzzle consisting of a pin, a hook, and a hasp, mounted on a wooden base (Fig. 1). The hasp was re-

Fig. 1. Mechanical puzzle whose solution consists of removing pin, disengaging hook from eye, and lifting hasp. (After Harlow *et al.*, 1950.)

strained by the hook which in turn was restrained by the pin. To disassemble the puzzle, it was necessary to disengage each device in serial order. A replica of the puzzle was attached to the home cage of each monkey throughout the experimental period. For subjects in group A, the puzzles were assembled and frequent checks were made for 12 consecutive days to see if one or more of the devices had been disengaged. If so, the puzzle was reassembled. For animals in group B, the experimenters never assembled the puzzles during this period and there was no evidence that the monkeys attempted to do so. On days 13 and 14, the puzzles were assembled for both groups and observations were made on each subject during five 5-minute periods, spaced 1 hour apart. Group A animals repeatedly disassembled the puzzle, usually within 60 seconds after it had been put together. Animals in group B, on the other hand, seldom solved the puzzle, and when solution did occur it never took place promptly.

This experiment was followed soon by another in which the performance of rhesus monkeys on a more complicated (six devices) puzzle was studied (Harlow, 1950). Again the puzzle was attached to the home cage. With the puzzle assembled, five 5-minute observations, 2 hours apart, were made daily. A correct response was defined as a device-opening and an error defined as touching any device out of sequence or touching the appropriate device yet failing to disengage it. Frequency of correct responses clearly increased and the error frequency decreased over the 12-day period. On day 13 the puzzles were checked every 6 minutes over a 10-hour interval, the number of devices opened was recorded, and the puzzles were immediately reassembled. The results testified to the persistence of manipulative behavior, as the experimenter found one or more devices open nearly every time he checked the puzzles. The total number of devices opened decreased as the session progressed, but this finding should not obscure the fact that the monkeys continued to respond to the puzzle for hours on end.

2. Discrimination Learning

Harlow and McClearn (1954) extended work on manipulative rewards to a more standard laboratory situation, a discrimination-learning problem. Rhesus monkeys were tested on a total of seven color-discrimination problems. The manipulanda consisted of five pairs of screw eyes, vertically arranged and attached to a board (Fig. 2). One member of each pair could be easily removed; the other was secured. The only clue to whether a screw eye was removable was its color. For example, in the first problem one member of the pair was red and removable; the other was green and fixed. Grasping and removing the red screw eyes was defined as a correct response. An incorrect response was defined as failing to remove a red screw eye if touched, or touching a green screw eye. Reward for a correct response was the opportunity to handle and closely examine the red screw eye for a brief interval. Each animal was tested 4 days a week for 7 weeks, with a different color-discrimination problem used each week. Four trials a day were given on each of four tests boards differing in the positional arrangements of the pairs of screw eyes. The animals' performance progressively improved from day 1 through day 4 on the first five problems. Performance on problems 6 and 7, which were presented following a 2-week rest interval, failed to show day-to-day improvement although performance even on day 1 of each problem was reasonably proficient; about 70% of all responses were correct.

Fɪɢ. 2. Rhesus monkey working on a color-discrimination problem for manipulative rewards. (From Harlow & McClearn, 1954.)

3. Development of Manipulation

The development of manipulative activities has been studied very thoroughly. Mason *et al.* (1959) measured the responsiveness of 30 infant rhesus monkeys to a variety of manipulanda. Manipulative responses were made during the first week of life and increased in strength and frequency as the infants matured. Pulling strings and chains, for example, began between day 6 and day 10 and increased greatly during the ensuing weeks. The genesis and development of knob manipulation followed a comparable course. Manipulative rewards can also promote learning in the infant monkey as well as the adult. Harlow *et al.* (1956) observed definite improvement in performance on a two-device mechanical puzzle and on a black-white discrimination-learning problem (see Chapter 11 by Zimmermann and Torrey). Again the only reward for correct responses was the opportunity to manipulate. Of theoretical significance was the fact that the subjects learned for manipulative rewards before they had handled solid foods. (See Section IV, A, 1 for further discussion.)

B. Visual Rewards

Anyone who spends a little time with rhesus monkeys knows that these animals are unflagging observers of environmental events. In the laboratory, numerous instances of visual-exploratory behavior are seen daily. For instance, in the Wisconsin General Test Apparatus (WGTA; see Chapter 1 by Meyer *et al.* in Volume I) an opaque screen is lowered during intertrial intervals so that the monkey cannot see the tester bait the foodwell. Unless prevented, a rhesus monkey will frequently lift the screen a short distance, enabling itself to watch the experimenter. Perhaps the simplest explanation for this kind of behavior is that the monkey is attempting to learn where the food will be found on the next trial. Yet, in a variety of situations where a barrier is inserted between the monkey and the experimenter, monkeys will devise almost ingenious ways to watch the actions of the experimenter. In some test conditions no food is ever visible. Observations such as these led directly to the notion that a rhesus monkey would learn simple tasks for the privilege of viewing activities in the environment.

1. DISCRIMINATION LEARNING

In the first of a series of experiments, rhesus monkeys were placed in a box about 2 feet on each side, illuminated from within by a 25-w lamp (Butler, 1953). At the front were two small doors with the inside face of each covered by a sheet of transparent plastic. Differently-colored cards could be inserted behind the plastic sheets. An opaque screen separated the doors from the monkey. Before each trial, the experimenter inserted a yellow card, for example, beneath the plastic of one door and a blue card beneath the plastic of the other door. Both doors were then closed, but the one holding the blue card was unlocked. To start a trial the experimenter raised the opaque screen, exposing the cards to the monkey. If the animal pushed against the door holding the blue card, the door opened for 30 seconds and the monkey was permitted to view the surroundings outside the box. In the original experiment, the small doors looked out on a room where several students usually congregated. Figure 3 shows a monkey soon after making a correct response on the color-discrimination problem. On completion of the reward interval the opaque screen was lowered. If the monkey pushed against the door holding the yellow card, a small light on the outside of the apparatus informed the experimenter of the monkey's error. The opaque screen was lowered immediately, preventing the animal from correcting itself. The intertrial interval was 30 seconds, and the position of the correct window was varied in accordance with a predetermined schedule.

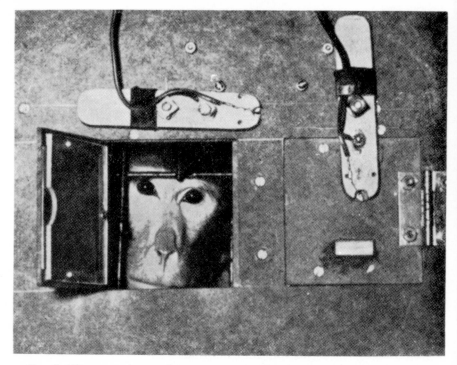

FIG. 3. Rhesus monkey performing in a visual-exploration box. (From Butler & Harlow, 1954.)

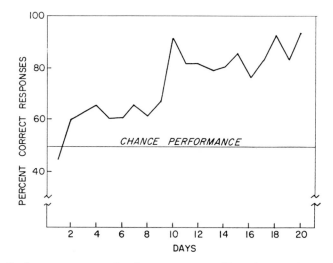

FIG. 4. Performance on a color-discrimination problem for visual reward. (After Butler, 1953.)

Figure 4 shows the mean performance of three monkeys on the yellow-blue discrimination problem. There is little question that learning occurred even though the animals did not reach a proficient level of performance for several days. The main point, however, is that merely a view of the events taking place outside the box was sufficient to maintain performance day after day. The relatively slow learning observed in this experiment was attributed to defects in apparatus design and the testing procedures. When, in a later study, the discriminanda were placed farther apart and the monkey was forced to wait for 5 seconds after presentation of the color stimuli before it could push against either door, color-discrimination performance improved greatly (Butler & Harlow, 1957).

2. SUSTAINED PERFORMANCE

Once it was shown that new responses could be learned using this technique of reward, the visually-rewarded behavior was tested for its persistence. It was found that monkeys would work several consecutive hours on color-discrimination problems with little or no evidence of being satiated by the visual rewards (Butler & Harlow, 1954). In a subsequent experiment, animals were tested 10 consecutive hours a day for 6 successive days (Butler & Alexander, 1955). During testing, the monkeys again were housed in an enclosed box. The box contained a single, small, spring-loaded door which could be easily opened by pressing against it. As soon as the monkey released the door it closed. The view from the box encompassed a colony of about 40 monkeys. Frequency of door-openings and the length of time that monkeys held the door open were recorded automatically. The subjects exhibited no day-to-day decline in total viewing time and the door was held open for a mean time of 4 hours daily! Moon's (1961) data also attest to the tenacity of visual-investigative behavior. In one of his experiments, rhesus monkeys were tested 2.5 hours a day for 10 consecutive days for a view of another monkey. Pressing a bar activated a motor-driven door which remained open as long as the animal held down the bar. The door remained open 50.8% of the time.

C. Auditory Rewards

Monkeys and apes of course attend to environmental sounds as well as sights. In the laboratory, monkeys vocalize vigorously when they hear the sounds associated with the preparation of their meals, become agitated when they hear the sounds of a fight in the colony, and become instantly still at the sound of a fear bark. Can sounds, like sights, be effective rewards in discrimination-learning problems? It turns out that rhesus monkeys will learn a simple task when the reward for a correct response is the opportunity to hear sounds made by other monkeys (Butler, 1957b). This was demonstrated by isolating a monkey in an enclosed box located in a sound-treated room. The box contained two levers and had a loudspeaker fixed to its top. A microphone and an audio amplifier were placed in an adjoining room, which housed a colony of monkeys. To ensure that the colony would be active vocally, testing preceded colony feeding. The monkey being tested, however, was always fed before the experimental session. If the test animal pressed one of the levers, it heard 12 seconds of sounds emanating from the colony. No extrinsic auditory stimuli followed a response to the other lever. More responses were made to the lever which provided the sounds of the other monkeys. Later, when the formerly "silent" lever was made the "reward" lever and the previous "reward" lever was made "silent," the subjects' pattern of responses shifted accordingly (Fig. 5). Frequency of responses did not decrease from one test session to the next.

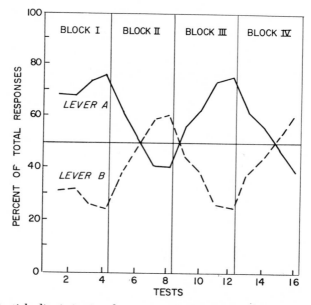

Fig. 5. Spatial discrimination learning and discrimination-reversal learning for auditory rewards. Lever A provided sound reward on test blocks I and III; lever B provided sound reward on test blocks II and IV. (After Butler, 1957b.)

D. Comments on Performances for Investigatable Rewards

The formal experiments described above (Sections II, A, B, C) form the basis for most of the subsequent studies of investigative behavior in nonhuman primates. They have in common the feature that receipt of investigatable rewards was contingent upon the execution of a learned discriminative response. Monkeys learned a color-discrimination problem in one situation for manipulative rewards and in another for visual rewards. A spatial discrimination problem was learned for auditory rewards. In the mechanical puzzles, monkeys were required to learn, on the basis of the current spatial arrangements of the devices, which ones were movable. It is assumed that movability of the devices, not puzzle solution, served as the reward. The work on infant monkeys shows that a strong manipulative propensity is present by the time an animal is 2 to 3 weeks old. The studies that follow were designed to delineate the conditions that increase and those that decrease the amount of investigative activities.

III. DETERMINANTS OF INVESTIGATIVE BEHAVIOR

A. Variables that Influence Manipulative Activities

Unfamiliar stimuli in a familiar setting are likely to excite the interest of animals. They exhibit either curiosity or fear. This commonplace observation led McDougall (1923, pp. 152–153) to postulate that these two "instincts" are the antitheses of one another. In the recent experimental literature, degree of novelty has been implicated as an important determinant of an animal's reaction to a strange situation. Several studies with chimpanzees demonstrate clearly the role of novelty. Welker (1956a) presented 15 pairs of objects sequentially to a group of chimpanzees. One member of each pair differed from the other in only one or two ways. One, for example, would be brighter, or a different color, or rougher, or movable, or of a different shape. In some pairs, manipulation of one of the members would produce a sound or turn off a light. The animals were tested usually once a day with the session lasting for 6 minutes. Manipulation and visual orientation toward the objects were recorded. Invariably, as the animals gained more experience with a pair of objects, the frequency of manipulation and visual inspection decreased. When responsiveness appeared to reach an asymptotic level, another pair of objects was introduced. This procedure continued until the subjects had been exposed to all the object pairs. The total amount of manipulation and visual attentiveness decreased both within and between test sessions. Presentation of new object pairs, however, brought about an increase in investigative activities. Stimulus preferences were apparent, with animals attending more to objects that were brighter, larger, movable, more heterogeneous, and which provided additional light and sound stimulation. Welker (1956b) then chose heterogeneity for further study. He argued that heterogeneity made for increased novelty and hence such objects would receive more attention than would more homogeneous objects. To test this hypothesis, three sets of objects were used, each set consisting of 10 wood blocks. In one set, the blocks were identical in shape, size, and color. In the second set, the blocks differed from one another only with respect to color. Blocks in the third set differed with respect to color and/or form. The chimpanzees were presented an object set once a day for 9 days. Test sessions lasted for 10 minutes and manipulative behavior was recorded. The most heterogeneous set elicited the largest number of manipulative responses; more objects were touched and response was shifted from one object to another more frequently.

In Welker's first study, the younger animals (3 to 4 years) manipulated more than the older ones (7 to 8 years). Welker (1956c) next sought to find out how responsive still-younger chimpanzees were in such situations. He tested eight animals whose ages ranged from 10 to 57 months. The procedure was similar to that of his other studies. There were 10 stimulus situations; each consisted of one, two, or three objects with the size and shape of the objects varying from one situation to the next. An animal was exposed to a situation daily for 5 minutes. The youngest chimpanzees withdrew from the objects and watched from the opposite end of the cage. After a few exposures, they approached and touched the objects. Soon they manipulated them freely whenever presented. When a new set of objects was introduced, however, the younger animals again retreated and only after several sessions began examining the objects manually. The 3- and 4-year-olds, on the other hand, started manipulating the objects immediately after presentation.

Working within the framework provided by Welker, Menzel *et al.* (1961) showed elegantly that the degree of novelty affects the young chimpanzee's manipulative behavior. Two animals, 25 and 27 months old, served as subjects. They had been maintained in isolation for the the first 21 months of life and their experience with novel stimuli had been controlled. The manipulanda consisted of eight wood blocks that differed in size (1.25 or 2.25 inches), color (black or white), and shape (cube or triangle). A small white cube was placed in each animal's living quarters several days before formal testing began, remaining there throughout the study except during the time the animals were being tested. This object was the "familiar" object from which the others differed in one, two, or three ways. Eight 60-second exposures to each object were given daily with a record kept of the length of time an object was contacted. The degree of novelty clearly influenced the chimpanzees' performance; mean contact duration increased directly with the number of novel cues.

In a second experiment, Menzel *et al.* traced the orderly development of approach to strange objects. Fifty small objects differing from one another in a great variety of ways were presented sequentially to the chimpanzees. Four 5-minute trials were given daily. As soon as an animal showed no fear of the object and manipulated it freely, another object was introduced. After having experience with many different objects, the two infants approached and manipulated the objects almost as freely on the first trial as they did on subsequent ones. Figure 6 illustrates what may be construed as the gradual ascendancy of an approach over withdrawal reaction to new objects.

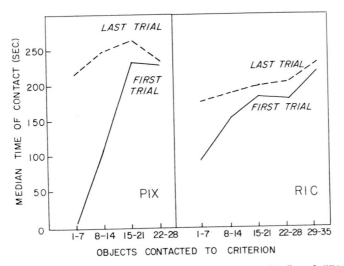

Fig. 6. Gradual reduction of reluctance of chimpanzees "Pix" and "Ric" to approach and manipulate new objects. (After Menzel et al., 1961.)

Those data available on the role of novelty in maintaining manipulative behavior in monkeys correspond closely to the chimpanzee material. New objects are handled much more frequently than familiar objects; objects that are visually conspicuous elicit more response than objects with few distinctive stimulus characteristics. Three experiments demonstrate these phenomena. Carr and Brown (1959a) placed rhesus monkeys in a box where three different objects, attached to the wall, were accessible for manipulation. These objects differed in size and shape, but for any given test session they were of the same material: wood, rubber, cork, wire, plastic, or metal. Subjects were presented objects of identical material for five consecutive daily tests; then new objects were introduced. Each test session lasted for 10 minutes, and the number of manipulations as well as the total time spent manipulating were recorded automatically. The results showed that for each class of objects, manipulative activity declined from day to day. When an object of a different material was substituted, the amount of manipulation increased. Monkeys were most responsive to wood objects and somewhat less so to rubber and cork objects. Carr and Brown (1959b) reported other data which suggest that if all objects are visually identical, the amount of manipulation is not influenced significantly by the objects' composition.

Mason (1961) recorded the manipulative responses of two rhesus monkeys from shortly after birth until they were 3 months old. Each infant was housed in a cage partitioned into three compartments by horizontal bars. The subject was confined to the center compartment and a stimulus object was placed within easy reach in one of the end compartments. Six stimulus objects were used: two were movable; two contained a flashing light; two were stationary. All differed in size and shape. An object was presented for 5 consecutive days, remaining in one compartment for 3 days and in the other for 2 days. Discrete contacts with an object were recorded automatically. Each object was presented for three 5-day periods with 30 days intervening before the same object was shown again. As in the earlier work on infant monkeys (Section II, A, 3), the amount of manipulation increased strikingly, particularly during the first 5 weeks of life. Daily exposure to the same object decreased responsiveness, which returned to a higher level when a different object was introduced. Objects that were movable or contained the flashing light were contacted significantly more than the stationary objects.

B. Variables that Influence Visual Exploration

1. SHORT-TERM RESTRICTION OF VISUAL EXPERIENCE

In Butler and Alexander's (1955) study of sustained performance (Section II, B, 2), the frequency and duration of door openings were controlled by the subject. As the experiment progressed, response frequency decreased but response duration increased, the result being that total viewing time remained about the same day after day. It was suggested at the time that monkeys will work to attain a relatively fixed amount of daily visual exploratory experience by manipulating those dependent variables under their control. This suggestion implies that monkeys deprived of a varied visual environment will work more diligently for visual reward. This possibility has been examined. Rhesus monkeys were kept in an enclosed, but illuminated, box for 0, 2, 4, or 8 hours before being given the opportunity to visually survey the surroundings outside the box (Butler, 1957a). At the end of the deprivation period they were tested for 1 hour in which responses to the door were rewarded with a 12-second view of a monkey colony, in accordance with a variable-interval schedule of reinforcement. Response frequency increased as the deprivation period lengthened from 0 to 4 hours. The mean number of responses after 8 hours of visual restriction was not appreciably higher than that recorded after the 4-hour period.

In a recent study, Fox (1962) confined macaque monkeys (*Macaca assamensis* and *Macaca nemestrina*) in darkness for 0, 1, 2, 3, 4, or 8 hours. After the deprivation period, pressing a bar produced a 0.5-second flash of light. Each test session lasted for 1 hour. Frequency of lever-pressing responses increased with increases in length of visual deprivation time up to 4 hours. As in the study described above, 8 hours of deprivation did not further increase the effectiveness of the visual reward (Fig. 7). Fox carried out an additional experiment in

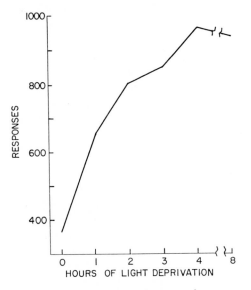

Fig. 7. Frequency of lever responses for light reinforcement as a function of length of light deprivation. (After Fox, 1962.)

which monkeys again were subjected to light deprivation ranging from 0 to 8 hours. During testing, however, a second lamp provided ambient illumination. Its intensity was ⅛, ¼, ½ or equal to that of the "reward" lamp. Response frequency decreased with increase in the level of ambient illumination under all conditions of deprivation. Wendt *et al.* (1963) kept a rhesus and a cynomolgus macaque (*Macaca irus*) in the dark 23 hours a day during the first 16 months of life. Later permitted to press a lever for light reward, the monkeys exhibited an extraordinarily high response rate. These studies demonstrate that restriction or deprivation of visual experience serves to increase the value of visual reward as measured by response frequency.

2. CHANGES OF REWARD SCENE

a. Drastic changes. What part did manipulative rewards play in the total performance of monkeys working for the opportunity to look outside the test box? The act of pushing open the door must have had some intrinsic reward value since manipulation of movable objects can effectively serve as a reward for learning new responses. Moreover, monkeys tested in this or a similar apparatus sometimes move the door back and forth repeatedly. And occasionally they will open the door, but look to the rear of the test box. Hence, early in the research program, it was deemed essential to find out whether or not the particular scene is in fact an important determinant of behavior in the visual-exploration box (Butler, 1954). Accordingly, rhesus monkeys were placed in an enclosed box which had only one door for viewing. On opening this door, the animals could look directly into a chamber which was either empty or contained (1) an assortment of food especially palatable to rhesus monkeys, (2) an electric train running on a circular track, or (3) another monkey confined to a small cage. Each response to the door provided 6 seconds of visual reward. Reliably more responses were made when the monkey or train was in the reward chamber than when it was empty or contained food. Assuming that the influence of the manipulative reward provided by the movable door was distributed equally among reward conditions, these data furnish convincing evidence that responsiveness can be markedly affected by the visual scene.

Rabedeau and Miles (1959) were also concerned about the contribution of manipulative rewards in a test situation where a view of the outside environment is contingent upon opening a door. They placed a visual-exploration box, similar to that used by Butler in a room that housed an assortment of equipment generally found in a behavior laboratory. Rhesus monkeys could attain a maximum view of 5 seconds for each response to the door, and frequency of door opening was recorded during five 24-minute tests. About a month later, the tests were rerun, but this time the overhead lamp in the room was turned off and illumination from other sources was blocked. With the room illuminated, response frequencies were reliably greater, yet the subjects did continue to push open the door when the room was dark.

It appears, then, that both visual and manipulative rewards influence behavior in the type of apparatus employed ostensibly to study responsiveness for visual rewards. However, visual rewards can clearly be made predominant. A dramatic example of this is illustrated in Fig. 8. Here, a monkey, placed in an enclosed box, was put on a variable-interval schedule of visual reward. For the first hour, the viewing door faced a monkey colony; for the next it faced a homogeneous white screen. On the second test day, the order of presentation was reversed. The response rate was strikingly elevated when a colony of monkeys could be seen (Butler, 1960, p. 169).

One could argue that an "exciting" visual reward could markedly raise the general activity level. In the visual-exploration box, an increase in responsiveness to visual stimuli could be attributable largely to an increase in other activities, notably manipulation. This idea, although credible, is not supported by the data. Symmes (1959) provided monkeys with both a door to open and a small hole to peer through. The door opened only on a small dark cubicle, while the peephole afforded a view of the happenings outside the box. The two devices were sufficiently close together that an animal could simultaneously look through

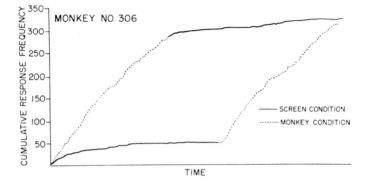

FIG. 8. Changes in responsiveness associated with changes in the visual reward. (After Butler, 1960.)

the aperture and manipulate the door. When a series of slides was projected on a screen outside the box, the time spent peeping through the aperture increased reliably over that recorded when the screen was blank. The frequency of manipulative responses to the door, on the other hand, remained quite stable throughout the period in which conditions of visual reward were changing.

b. Subtle changes. Once it is shown that responding for visual rewards can be influenced by varying the reward scene drastically, the effects of more subtle changes can profitably be investigated. By means of the paired-comparison technique, some progress in this direction has been made. Menzel and Davenport (1961) permitted four chimpanzees to view a human being from either of two plastic windows. One window afforded a clear view, but the plastic in the other was bent so that the perceived images were, for the human observer, grossly distorted. The subjects spent more time looking through the window providing a clear view. The chimpanzees sampled both windows several times before settling down to the one that did not produce distortion.

This line of research has also been followed with rhesus monkeys as subjects (Butler & Woolpy, 1963). Animals were placed in a box containing two small windows. Each window faced a projection screen as shown in Fig. 9. During a test session the projection equipment operated continuously, but the projection lamp was lit only when a monkey approached within 1.5 inches of one of the windows. In other words, a monkey going to one window could view one slide or motion picture, and could see a different one by going to the opposite window. As soon

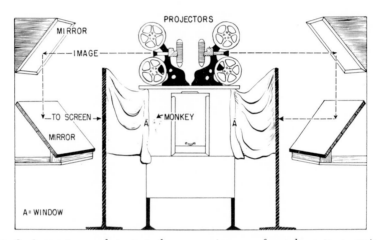

Fig. 9. Apparatus used to test the responsiveness of monkeys to a variety of projected stimuli. "A" represents the location of the windows. (From Butler & Woolpy, 1963.)

as the animal withdrew from a window the projection lamp automatically went out. The frequency and duration of response to each window were registered. It should be noted that viewing the slides or motion pictures was not contingent on manipulative responses; an animal was required only to position its head appropriately. Rhesus monkeys, like chimpanzees, preferred to view a scene (motion picture of other animals) that was clearly focused. Symmes' (1959) findings on the reward value of projected slides were also confirmed. That is, monkeys spent no more time viewing a single projected slide than a blank screen, but a series of slides sequentially projected increased their responsiveness. Aside from corroborating the results of others, a considerable amount of new information was obtained with this use of projected stimuli. When monkeys were given a choice of viewing a series of automatically changing slides or a motion picture, they overwhelmingly chose the latter. In addition, the monkeys viewed longer when motion pictures were projected brightly rather than dimly, at normal speeds rather than at slower speeds, in color rather than in black and white, and right side up instead of upside down. For each of these reward conditions, two copies of the same film were projected simultaneously; they differed only with respect to one of the characteristics listed above.

Like the chimpanzees in the two-choice situation, the rhesus monkeys sampled both windows about the same number of times. The preference for one visual reward over the other was expressed primarily by the length of time a subject remained at a window.

C. Social Rewards

Several of the studies already described employed the sights and sounds of other monkeys as rewards for acquiring new responses. As a class, these rewards are probably the most effective in maintaining responsiveness at a high level. However, some of these rewards are reliably more effective than others. For example, monkeys pressed a lever more frequently to hear the call of an isolated monkey than to hear the vocalizations of a highly agitated monkey colony (Butler, 1958). No data are available on whether a monkey will work more readily to hear the calls of one particular monkey than to hear the calls of another, but some animals apparently prefer to see one individual rather than another. Unpublished observations by the writer suggest that, in a two-choice viewing situation, a monkey will usually spend more time looking at its cagemate than viewing another monkey in the colony.[1] Furthermore, the compatibility of cagemates and the length of time the two have been housed together make a difference. Sex, too, can play a decisive role. For example, male monkeys generally prefer to view the female cagemate over others that they see daily in the colony room. But they are apt to prefer a strange female to their cagemate, especially if the stranger is in full sexual coloration.

Harlow and Zimmermann (1958) used a modified visual-exploration box to illustrate the rewarding effect on infant monkeys of surrogate mothers that provided contact comfort. On pressing a lever the infants could view either a cloth surrogate, a wire surrogate, or an empty cage. Reliably more responses occurred when they were rewarded by a view of the cloth surrogate. Usually the infants were at least 10 days old before being placed in the visual-exploration box. Harlow and Zimmermann

[1] William A. Mason (personal communication) also has evidence that a cagemate is more rewarding than a "stranger." When one member of the pair is confined to a small cage, the other will frequently pull a chain attached to that cage which opens a door, releasing the inmate. Releases are far more numerous among cagemates than among monkeys housed individually.

noted however, that even 3-day-old infants exhibit a "compulsive visual curiosity."

D. Suppression of Investigative Behavior

When the role of novelty in eliciting manipulative responses was discussed (Section III, A), the point was made that new or strange objects can evoke both curiosity (approach) and fear (withdrawal). Frequently, animals are first fearful of new things, but with repeated exposure fear subsides and active, investigative responses predominate. Hebb (1946) studied the reactions of chimpanzees to visual objects chosen primarily for their fear-provoking properties. A chimpanzee was induced to come to the front of its living cage by an offer of food. While it was there, an object concealed in a box was presented by opening the box. Fear was inferred if the animal withdrew from the exposed object. Vastly different objects were used; some were primate objects such as an adult-chimpanzee head made of papier-mâché and an unclothed doll representing a human infant; some were nonprimate objects, for example, a mechanical "grasshopper" and a rubber dog; still others were pictures chosen for their neutral affective value. The most effective fear-provoking stimuli were a chimpanzee skull with moving jaw, a painted wax snake, and a mounted skin of a spider monkey whose head and shoulders moved. Among the least effective were a dead infant chimpanzee, a small wax grub, and a moving rubber tube ("snake"). As the experiment progressed, investigative behavior differed considerably among animals, some showing fewer withdrawal responses and others retreating to the rear of the cage whenever the box concealing the objects appeared. In less formal observations, it was noted that the sight of an anesthetized adult chimpanzee elicited strong fear in the other apes. Hebb thought that the fear of mutilated and inert primate bodies was spontaneous. In the context of this chapter, these visual experiences served to suppress investigative behavior in many of the animals.

Hebb's observations provided the impetus for finding out what kinds of experiences, auditory as well as visual, provoke fear in monkeys to the extent that investigative behavior is severely curtailed (Butler, 1964). Several different kinds of stimuli were selected. Some were comparable to those Hebb found to be particularly distressing to chimpanzees, viz., an anesthetized adult monkey and a perfused adult monkey with its head placed in its hands. Another was a view of a monkey confined to a small box containing three live snakes, with the snakes out of sight of the observing monkey. The monkey that served as the visual object appeared to the human observer to be most fearful. During testing, the monkeys could view these scenes by merely approaching a small window. Surprisingly enough, there was no clear-cut evidence that viewing behavior was substantially suppressed by any of these visual experiences. Perhaps the test box served psychologically as a sanctuary for the subjects. While peering at the world outside they could withdraw from the window at any time. Symmes (1959) observed a comparable phenomenon. A large photograph of a frightened monkey, which appeared to induce fear when monkeys were required to push open a door to view it, failed to do so when monkeys could view it through a peephole. Symmes believed that the monkeys were reluctant to expose themselves in fear-provoking situations; the more "daring" response of opening a door was suppressed.

Stimuli can be found that will suppress investigative behavior in the monkey even when the animal is operating under optimally "safe" conditions. Observing responses through a peephole drop precipitously when they lead to a close-up view of a live snake.

Even with live snakes, the experimental situation can be contrived so that monkeys will work diligently to see them. Using as background the report of Moon and Lodahl (1956) that monkeys will perform a simple response when rewarded by an increase or decrease in illumination, Butler (1964) placed a monkey and live snake together in a totally dark box for an extended series of 10-minute test sessions. Pulling a chain suspended from the top of the box turned on a lamp for 6 seconds. Response frequencies were reliably greater when the snake was present. When pulling the chain turned the light off rather than on, response frequencies dropped dramatically. If we can assume, as some of the literature suggests, that monkeys fear snakes, it would follow that they would tend to avoid snakes. To do this in the situation described, they had to be able to see the snake; hence chain-pulling reached a high level when it served to keep the box illuminated. Yet, increased fear suppresses behavior even in this emergency-type situation. The data supporting this statement were obtained when, 3 minutes after the start of the test, either recorded alarm cries of monkeys or interrupted white noise was played for 15 seconds. Chain-pulling responses immediately decreased (Fig. 10). When the sounds were alarm cries, most monkeys seldom pulled the chain for the remainder of the session.

IV. DISCUSSION

A. Theories of Investigative Behavior

Curiosity-investigative motives have been postulated to account for the responsiveness of nonhuman primates to manipulative, visual, and auditory rewards. Harlow (1953b) has contended that these motives are innate and not conditioned upon the so-called primary biological drives such as hunger, thirst, and pain. The interpretation, however, has engendered strong and varied opposition. The main objection has been that there is not enough experimental evidence to warrant postulating a new motive; these behaviors can be explained adequately in terms of existing behavior theories.

1. SECONDARY REINFORCEMENT

The concept of secondary reinforcement, for example, has been advanced to account for many of the experimental findings described in this chapter. The logic of this argument appears simple: reinforcement seemingly inherent in manipulation has been established through a long

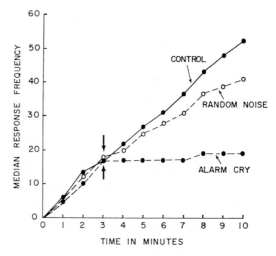

FIG. 10. The marked suppressive effect of a fear-provoking sound (alarm cry) on chain-pulling. The arrows indicate onset of auditory stimuli. No sounds were presented during the control tests. (From Butler, 1964.)

association between manipulating edible objects and eating (reduction of hunger). Further, the visual rewards used in the early studies usually were views of other monkeys and humans. According to the secondary-reinforcement thesis, the sight of a human could be associated with the reduction of hunger and thirst drives since the colony was routinely fed and watered by humans. A view of other monkeys is associated with a variety of sociosexual experiences which could form the basis for a complicated network of drive reductions. And the auditory rewards used in the original study by Butler (1957b) were those sounds that monkeys emit before eating; hence, it was certainly possible that they provided secondary reinforcement.

It seems to this writer that any explanation in terms of secondary reinforcement is most unsatisfactory. Monkeys perform manipulative and visual-exploratory tasks at the age of 1 to 2 weeks, before they are fed solid foods. To add to the difficulties, manipulation and visual exploration are extremely persistent, but no conditioned behavior other than that based on pain has been shown in the laboratory to endure unless a primary reinforcer is occasionally dispensed. Finally, it is more circuitous to say that investigative activities are reinforced by stimuli that have been associated with drive reduction than to say that they are reinforced because they cause some drive to be reduced.

2. Drive Reduction

The stimulation resulting from investigative activities might reduce a drive, thereby reinforcing such activities. Several drives have been proposed which might be reduced by this stimulation. In each case, if monotony or lack of contact with environmental events is drive-arousing, then auditory and tactile as well as visual exploration would reduce the proposed drive. Brown (1953; 1961, pp. 337–338) has suggested that anxiety reduction might be the mechanism underlying visual-exploratory behavior. He argued that being unable to see is a drive-arousing situation. The drive is reduced by the view obtained by pushing open a door; hence, the door-opening response (in later experiments the window-peeking response) is reinforced. Myers and Miller (1954) also mentioned that confinement might produce anxiety. Myers and Miller further suggested that investigative behavior might be an expression of a boredom drive aroused by monotonous surroundings, reduced by stimulus change. Stimulus satiation (Glanzer, 1953) is another proposed explanation, similar to the concept of boredom. According to this view, animals avoid stimuli to which they have been exposed.

The various drive-reduction explanations are identical in their logic. Each names an aversive drive state, reduction of which is supposed to be reinforcing. Each implies that investigative activities are escape or avoidance reactions. Calling the drive "curiosity" instead of "boredom" does not change this implication. By its nature, drive-reduction theory implies that *all* learned behavior is escape or avoidance responding. But perhaps investigative behavior is not learned behavior. Such a view is more parsimonious than explanations based on either secondary reinforcement or drive reduction.

3. Innate Investigative Responsiveness

The most direct explanation of investigative responsiveness is that such activities are innate rather than learned. In certain stimulus situations (e.g., peephole or novel object) the most probable response of the primate is to investigate (e.g., look or manipulate). In some experiments, however, learning has been demonstrated: monkeys learn which door or lever to press, or which part of a puzzle to release. The prepotent investigative responses themselves could reinforce such learned responses. There is no need to postulate a new behavioral principle to account for such reinforcement, since it can be predicted from at least one general hypothesis already available. Premack (1959) proposed that a reinforcer is any response that is more probable than the response to be reinforced. In addition to explaining the results presented in this chapter, the same hypothesis can also account for the effects usually offered as support for a drive-reduction theory, for example, running reinforced by drinking, and for effects not easily predicted from drive-reduction theory, for example, drinking reinforced by running (Premack, 1962). Premack (1963) has also shown that this reinforcement relation holds for manipulative responses of cebus monkeys. Considering certain pairs of manipulanda, one response occurred less frequently than another when the monkey could respond freely. Later, when the second response could not be performed until the first one had been made, the rate of the first response increased. This indicated that the second, more probable response had reinforced the first, less probable response. Thus, the data presented in this chapter can be explained by principles of very great generality: (a) innate responsiveness and (b) reinforcement of less probable responses by more probable responses. It is unnecessary to postulate additional, more complex mechanisms of secondary reinforcement or reduction of hypothetical drive states.

The importance of the data on investigative behavior is largely independent of the explanation advanced to account for them. The experiments demonstrate that behavior can be influenced in a consistent manner by a wide variety of rewards hitherto not used. After the original studies using manipulative, visual, and auditory rewards, the experiments were designed to provide more detailed information on the reward conditions which differentially influence behavior. This area of research is now being emphasized.

B. Functional Significance of Investigative Behavior

Questions concerning the functional significance of investigative behavior and its role in the evolution of behavior are more general and, to some students, more intriguing. Investigative behavior, when defined as an approach response to a change in the environment, has been shown to occur in animals as low in the phylogenetic scale as the cockroach (Darchen, 1952). Incidences of curiosity in birds are ubiquitous. Nissen (1951, pp. 357–359) mentioned that these behaviors are less prominent in nonmammalian species than in the mammals, especially the primates, but the propensity to investigate in no way distinguishes primates from other animals. The differences appear to be the degree to which a primate can act upon its environment and its capacity to perceive and be guided by the consequences of action. A young child, through continued visual, auditory, and manipulative examination of his surroundings, acquires an enormous amount of information; by the time he reaches 2 years of age he has gained mastery over a considerable portion of his world. Lower animals too become familiar with their environment though exploration, an attainment which unquestionably serves their maintenance and safety requirements. But monkeys and apes, at least, learn things which at the time appear to have no direct bearing on their physical welfare. For instance, when given the opportunity to handle sticks, chimpanzees learn to apply them to external objects; they poke at things and, with practice, connect two sticks together. Those animals that have had more experience manipulating sticks are generally better able to use them as tools to get food (Birch, 1945; Schiller, 1952). Klüver (1933, pp. 258–305), after describing in detail a cebus monkey's use of tools, suggested that through careful examination the monkey learns about the properties of objects and is able to "grasp the principles involved in the instrumental use of objects" (Klüver, 1933, p. 300). Harlow's rhesus monkeys, though not noted for their tool-using abilities, certainly learned the contingencies and constraints that were built into the mechanical puzzles. It requires little imagination to see how intrinsically-motivated learning of the physical relations between objects can be an asset in exerting some control over the environment.

Although monkeys and chimpanzees may be inveterate observers, one wonders whether they learn anything about the relations among events by merely watching. This question has been asked repeatedly since the beginning of animal experimentation in learning. The answer is that some primates unquestionably can learn by observing the actions of others. By observing the behavior of the experimenters, the celebrated chimpanzee, Viki, learned rapidly to push food from a tunnel using a stick, to operate a wall switch which caused food to drop from the ceiling lamp, and to push a suspended string with a stick, thus causing a box containing food to open (Hayes & Hayes, 1952). Of the many claims that monkeys can learn by observing, that of Darby and Riopelle (1959) is the most convincing. In their experiment, one rhesus monkey observed another perform a long series of two-choice visual discrimination problems for food reward. The demonstrator would make a choice, the discriminanda would be returned to their original positions, and then the observer would be permitted to respond. The performance of the observer improved as the experiment progressed. What is exciting about this experiment is that the observer obtained more information about a particular problem when the demonstrator made an incorrect response than when it responded correctly. This means that the observing monkey was not merely repeating the acts of the demonstrator; it was responding to the consequences of the demonstrator's behavior (see Chapter 2 by Miles in Volume I).

In conclusion, the experimental data summarized in this chapter indicate that monkeys and apes strive to maintain visual contact with environmental events. When the opportunity is provided, they will reach for manipulable things, usually choosing those that are visually more conspicuous, and examine them keenly. G. Elliot Smith (1927), in his discussion of man's evolution, attributed profound significance to this intimate association between visual and tactile exploration. "Man has evolved as the result of the continuous exploitation throughout the Tertiary Period of the vast possibilities which the reliance upon vision as the guiding sense created for a mammal that had not lost the plasticity of its hands by too early specialization. Under the guidance of vision the hands were able to acquire skill in action and incidentally to become the instruments of an increasingly sensitive tactile discrimination, which again reacted upon the motor mechanisms and made possible the attainment of yet higher degrees of muscular skill. But this in turn reacted upon the control of ocular movements and prepared the way for the acquisition of stereoscopic vision and a fuller understanding of the world and the nature of the things and activities in it" (Smith, 1927, p. 152).

REFERENCES

Berlyne, D. E. (1950). Novelty and curiosity as determinants of exploratory behavior. *Brit. J. Psychol.* **41**, 68.

Birch, H. G. (1945). The relation of previous experience to insightful problem-solving. *J. comp. Psychol.* **38**, 367.

Brown, J. S. (1953). Comments on Professor Harlow's paper. *In* "Current Theory and Research in Motivation" (J. S. Brown, H. F. Harlow, L. J. Postman, V. Nowlis, T. M. Newcomb, & O. H. Mowrer, eds.), pp. 49–55. Univer. Nebraska Press, Lincoln, Nebraska.

Brown, J. S. (1961). "The Motivation of Behavior." McGraw-Hill, New York.

Butler, R. A. (1953). Discrimination learning by rhesus monkeys to visual-exploration motivation. *J. comp. physiol. Psychol.* **46**, 95.

Butler, R. A. (1954). Incentive conditions which influence visual exploration. *J. exp. Psychol.* **48**, 19.

Butler, R. A. (1957a). The effect of deprivation of visual incentives on visual exploration motivation in monkeys. *J. comp. physiol. Psychol.* **50**, 177.

Butler, R. A. (1957b). Discrimination learning by rhesus monkeys to auditory incentives. *J. comp. physiol. Psychol.* **50**, 239.

Butler, R. A. (1958). The differential effect of visual and auditory incentives on the performance of monkeys. *Amer. J. Psychol.* **71**, 591.

Butler, R. A. (1960). Acquired drives and the curiosity-investigative motives. *In* "Principles of Comparative Psychology" (R. H. Waters, D. A. Rethlingshafer, & W. E. Caldwell, eds.), pp. 144–176. McGraw-Hill, New York.

Butler, R. A. (1964). The reactions of rhesus monkeys to fear-provoking stimuli. *J. genet. Psychol.* **104**, 321.

Butler, R. A., & Alexander, H. M. (1955). Daily patterns of visual exploratory behavior in monkeys. *J. comp. physiol. Psychol.* **48**, 247.

Butler, R. A., & Harlow, H. F. (1954). Persistence of visual exploration in monkeys. *J. comp. physiol. Psychol.* **47**, 258.

Butler, R. A., & Harlow, H. F. (1957). Discrimination learning and learning sets to visual exploration incentives. *J. gen. Psychol.* **57**, 257.

Butler, R. A., & Woolpy, J. H. (1963). Visual attention in the rhesus monkey. *J. comp. physiol. Psychol.* **56**, 324.

Carr, R. M., & Brown, W. L. (1959a). The effect of the introduction of novel stimuli upon manipulation in rhesus monkeys. *J. genet. Psychol.* **94**, 107.

Carr, R. M., & Brown, W. L. (1959b). Manipulation of visually homogeneous stimulus objects. *J. genet. Psychol.* **95**, 245.

Darby, C. L., & Riopelle, A. J. (1959). Observational learning in the rhesus monkey. *J. comp. physiol. Psychol.* **52**, 94.

Darchen, R. (1952). Sur l'activite exploratrice de *Blatella germanica*. *Z. Tierpsychol.* **9**, 362.

Dashiell, J. F. (1925). A quantitative demonstration of animal drive. *J. comp. Psychol.* **5**, 205.

Dennis, W. (1955). Early recognition of the manipulative drive in monkeys. *Brit. J. anim. Behav.* **3**, 71.

Dennis, W., & Sollenberger, R. T. (1934). Negative adaptation in the maze exploration of albino rats. *J. comp. Psychol.* **18**, 197.

Fitzgerald, Alice (1935). Rearing marmosets in captivity. *J. Mammal.* **16**, 181.

Fox, S. S. (1962). Self-maintained sensory input and sensory deprivation in monkeys: A behavioral and neuropharmacological study. *J. comp. physiol. Psychol.* **55**, 438.

Girdner, J. B. (1953). An experimental analysis of the behavioral effects of a perceptual consequence unrelated to organic drive states. *Amer. Psychologist* **8**, 354. (Abstract.)

Glanzer, M. (1953). Stimulus satiation: An explanation of spontaneous alternation and related phenomena. *Psychol. Rev.* **60**, 257.

Harlow, H. F. (1950). Learning and satiation of response in intrinsically motivated complex puzzle performance by monkeys. *J. comp. physiol. Psychol.* **43**, 289.

Harlow, H. F. (1953a). Mice, monkeys, men, and motives. *Psychol. Rev.* **60**, 23.

Harlow, H. F. (1953b). Motivation as a factor in the acquisition of new responses. *In* "Current Theory and Research in Motivation" (J. S. Brown, H. F. Harlow, L. J. Postman, V. Nowlis, T. M. Newcomb, & O. H. Mowrer, eds.), pp. 24–49. Univer. Nebraska Press, Lincoln, Nebraska.

Harlow, H. F., & McClearn, G. E. (1954). Object discrimination learned by monkeys on the basis of manipulation motives. *J. comp. physiol. Psychol.* **47**, 73.

Harlow, H. F., & Zimmermann, R. R. (1958). The development of affectional responses in infant monkeys. *Proc. Amer. phil. Soc.* **102**, 501.

Harlow, H. F., Harlow, Margaret K., & Meyer, D. R. (1950). Learning motivated by a manipulation drive. *J. exp. Psychol.* **40**, 228.

Harlow, H. F., Blazek, Nancy C., & McClearn, G. E. (1956). Manipulatory motivation in the infant rhesus monkey. *J. comp. physiol. Psychol.* **49**, 444.

Hayes, K. J., & Hayes, Catherine (1952). Imitation in a home-raised chimpanzee. *J. comp. physiol. Psychol.* **45**, 450.

Heathers, G. L. (1940). The avoidance of repetition of a maze reaction as a function of the time interval between trials. *J. Psychol.* **10**, 359.

Hebb, D. O. (1946). On the nature of fear. *Psychol. Rev.* **53**, 259.

Hobhouse, L. T. (1901). "Mind in Evolution." Macmillan, London.

Kellogg, W. N., & Kellogg, Laverne A. (1933). "The Ape and the Child." McGraw-Hill, New York.

Klüver, H. (1933). "Behavior Mechanisms in Monkeys." Univer. Chicago Press, Chicago, Illinois.

Kohts, Nadie (1935). "Infant Ape and Human Child." Museum Darwinianum, Moscow.

Lowther, F. deL. (1940). A study of the activities of a pair of *Galago senegalensis moholi* in captivity, including the birth and postnatal development of twins. *Zoologica* **25**, 433.

McDougall, W. (1923). "Outline of Psychology." Scribner's, New York.

Mason, W. A. (1961). Effects of age and stimulus characteristics on manipulatory responsiveness of monkeys raised in a restricted environment. *J. genet. Psychol.* **99**, 301.

Mason, W. A., Harlow, H. F., & Rueping, R. R. (1959). The development of manipulatory responsiveness in the infant rhesus monkey. *J. comp. physiol. Psychol.* **52**, 555.

Menzel, E. W., Jr., & Davenport, R. K., Jr. (1961). Preference for clear versus distorted viewing in the chimpanzee. *Science* **134**, 1531.

Menzel, E. W., Jr., Davenport, R. K., Jr., & Rogers, C. M. (1961). Some aspects of behavior toward novelty in young chimpanzees. *J. comp. physiol. Psychol.* **54**, 16.

Montgomery, K. C. (1952). A test of two explanations of spontaneous alternation. *J. comp. physiol. Psychol.* **45**, 287.

Moon, L. E. (1961). Visual exploration as reinforcement of conditioned bar-pressing responses of monkeys. *J. exp. Anal. Behav.* **4**, 119.

Moon, L. E., & Lodahl, T. M. (1956). The reinforcing effect of changes in illumination on lever-pressing in the monkey. *Amer. J. Psychol.* **69**, 288.

Myers, A. K., & Miller, N. E. (1954). Failure to find a learned drive based on hunger; evidence for learning motivated by "exploration." *J. comp. physiol. Psychol.* **47**, 428.

Nissen, H. W. (1930). A study of exploratory behavior in the white rat by means of the obstruction method. *J. genet. Psychol.* **37**, 361.

Nissen, H. W. (1951). Phylogenetic comparison. *In* "Handbook of Experimental Psychology" (S. S. Stevens, ed.), pp. 347–386. Wiley, New York.

Premack, D. (1959). Toward empirical behavior laws: I. Positive reinforcement. *Psychol. Rev.* **66**, 219.

Premack, D. (1962). Reversibility of the reinforcement relation. *Science* **136**, 255.

Premack, D. (1963). Rate differential reinforcement in monkey manipulation. *J. exp. Anal. Behav.* **6**, 81.

Rabedeau, R., & Miles, R. C. (1959). Response decrement in visual exploratory behavior. *J. comp. physiol. Psychol.* **52**, 364.

Revesz, G. (1925). Experimental study in abstraction in monkeys. *J. comp. Psychol.* **5**, 293.

Romanes, G. J. (1892). "Animal Intelligence." Appleton, New York.

Schiller, P. H. (1952). Innate constituents of complex responses in primates. *Psychol. Rev.* **59**, 177.

Small, W. S. (1900). Experimental study of the mental processes of the rat. II. *Amer. J. Psychol.* **12**, 206.

Smith, G. E. (1927). "The Evolution of Man," 2nd ed. Oxford Univer. Press, London.

Symmes, D. (1959). Anxiety reduction and novelty as goals of visual exploration by monkeys. *J. genet. Psychol.* **94**, 181.

Thorndike, E. L. (1901). The mental life of the monkeys. *Psychol. Monogr.* **3**, No. 5 (Whole No. 15).

Welker, W. I. (1956a). Some determinants of play and exploration in chimpanzees. *J. comp. physiol. Psychol.* **49**, 84.

Welker, W. I. (1956b). Variability of play and exploratory behavior in chimpanzees. *J. comp. physiol. Psychol.* **49**, 181.

Welker, W. I. (1956c). Effects of age and experience on play and exploration of young chimpanzees. *J. comp. physiol. Psychol.* **49**, 223.

Wendt, R. H., Lindsley, D. F., Adey, W. R., & Fox, S. S. (1963). Self-maintained visual stimulation in monkeys after long-term visual deprivation. *Science* **139**, 336.

Yerkes, R. M. (1927). The mind of a gorilla. *Genet. Psychol. Monogr.* **2**, 1.

Yerkes, R. M., & Petrunkevitch, A. (1925). Studies of chimpanzee vision by Ladygin-Kohts. *J. comp. Psychol.* **5**, 99.

Chapter 14

The Radiation Syndrome[1]

Roger T. Davis

Department of Psychology, University of South Dakota, Vermillion, South Dakota

I. INTRODUCTION

Three special conditions have made it feasible and desirable to use nonhuman primates in studies of the psychobiological effects of ionizing radiation. First, there is evidence that the primate brain is more sensitive to damage by radiation than is the rat brain (Lindgren, 1958); second, research has been well subsidized because of urgent practical questions about the potential effectiveness of any human survivors of thermonuclear war; and third, chimpanzees (*Pan*) and rhesus monkeys (*Macaca mulatta*) have broad repertoires of responses and sensory equipment resembling man's.

A comprehensive account of the psychological literature that has grown up within radiation biology is beyond the scope of this chapter. Fortunately, two excellent reviews are available (Furchtgott, 1956, 1963). Any of several standard references in radiation biology can also be consulted for background (Claus, 1958; Hollaender, 1954–1956; Lea, 1955).

[1]This chapter was written while I was at the University of Oregon, on leave of absence from the University of South Dakota during 1962-1963. South Dakota researches were supported in part by the Division of Radiological Health, Bureau of State Services, Public Health Service, Grant RH66-05. I am very much indebted to R. W. Leary and A. A. McDowell for their helpful suggestions, but they should not be held responsible for my conclusions.

Irradiation of animals involves bombarding all or part of the body with either subatomic particles or a short-wave portion of the electromagnetic spectrum. Either type of radiation dissociates molecules into their ionic components. Free radicals, thus formed, are extremely reactive chemically and the recombination of radicals into new compounds is unfavorable to cellular life.

The unit of dosimetry for X- or γ-radiation is the roentgen (r), which is defined as "the quantity of X- or γ-radiation such that the associated corpuscular emission per 0.001293 g. of air produces, in air, ions carrying 1 electrostatic unit of quantity of electricity of either sign" (Lea, 1955, pp. 6–7). The biological consequences of irradiation with particles are greater than those of irradiation with the same number of roentgens of X- or γ-radiation. Therefore, an estimated factor of relative biological effectiveness (RBE) is multiplied by the amount of radiation to achieve the "roentgen equivalent mammal" (rem). Lethality of the effects of radiation is usually expressed as the percentage of subjects dying within a given period of time. The median lethal dose in a 30-day period is written LD-50/30.

The susceptibility of cells to the effects of ionizing radiations varies considerably. Cells that are dividing rapidly are as much as 1,000 times as susceptible to injury as cells that are not dividing. The central nervous system is very likely to be damaged during its formation, but thereafter the nervous system of the developing individual becomes decreasingly sensitive to the effects of radiation until ultimately there are no immediate histological changes with doses that are double or triple the LD-50 for whole-body radiation in the adult.

It is sometimes suggested that experiments dealing with the effects of radiation on behavior should employ only doses of radiation that would surely damage the central nervous system. Implicit in this suggestion is the assumption that histological stains reflect all possible changes in the structure and physiology of cells. Furthermore, the suggestion overlooks the possible effects on behavior of irradiating nonnervous tissue and the possible use of ionizing radiation as a stimulus.

Behavior has been demonstrably affected by radiation applied to all or part of an animal, in doses ranging from less than 1 r to more than 10,000 r, at various developmental stages from germ cell to maturity. Doses of radiation that are too low to cause tissue damage can act as stimuli. Low doses of radiation have served as unconditioned stimuli in conditioning avoidance of previously preferred liquids (Garcia *et al.*, 1961). The phenomenon has been demonstrated with a variety of mammals, including rhesus monkeys (Harlow, 1962). But nearly all studies employing low doses of radiation have used rats as subjects, as have studies of genetic and developmental effects of radiation.

Unlike the work with rodents, research with primates has been largely concerned with the chronic effects of irradiating adult animals. Most behavioral studies have examined either the effects of whole-body doses of radiation high enough to kill nearly half of the cases (LD-50), or those of irradiating the brain with doses high enough to damage neurons. The effects of these two conditions of irradiation will be discussed separately.

II. EFFECTS OF WHOLE-BODY IRRADIATION

Following a nuclear disaster many persons who had been irradiated would be called upon to perform medical and defensive services. Therefore, it is important to military and public-health authorities to ascertain the potentialities of persons irradiated in the range where immediate death is unlikely but death within 30 days is frequent. Consequently, federal support was made available to psychologists for studying this problem with nonhuman primates.

The effects of whole-body irradiation on the behavior of primates were studied extensively and concurrently at six different laboratories from 1952 to 1955: University of Wisconsin (Harlow & Moon, 1956), University of South Dakota (Davis et al., 1956), Emory University (Riopelle et al., 1956), University of Texas (Brown & McDowell, 1962a; Davis et al., 1958a), Yerkes Laboratories (Riopelle, 1962), and University of Washington (Leary & Ruch, 1955; Ruch et al., 1962). At the time these experiments began there was little information to indicate what measure of performance would best assay the effects of ionizing radiation on behavior. Therefore, batteries of tests were used at all six laboratories on the assumption that intertask interference was a less serious problem than the failure to select an appropriate measure.

The subjects were rhesus monkeys, except at Yerkes Laboratories which used chimpanzees. X-radiation from therapy machines was used at four laboratories, and mixed gamma and neutron radiation at Yerkes and Texas. Radiation was administered in single doses ranging from 350 to 400 r at Washington, Yerkes, Emory, and South Dakota. At South Dakota, each monkey received additional doses of 400 r and 300 r 14 months and 35 months after the first 400-r irradiation. At Wisconsin, each monkey received 100 r every 35 days until it died. A similar method was used with a group of monkeys at Emory, some of which received 100 r every 2 weeks while others received 200 r every 2 weeks until 1,000- and 2,000-r doses were accumulated, respectively. The results with the latter dose were unlike the finding at Wisconsin that monkeys die with similarly accumulated doses around 1,200 r. Monkeys at Texas received 139- to 1,116-rem doses (assuming the RBE for neutrons is 10) divided either into 20 exposures spaced 12 days apart or 40 exposures spaced

4 days apart. In spite of variation in procedures from laboratory to laboratory, the methods of training the subjects and the techniques of irradiation were similar. The most noticeable fault in procedure was that of Riopelle *et al.* (1956), who irradiated subjects in a metal cage which probably caused more scattering of radiation than a plastic cage would have.

Two other studies used rather different techniques of whole-body irradiation. One, which involved the same range of doses as the above studies, was conducted with rhesus monkeys irradiated in the field at the Nevada Test Site with an exploding nuclear device as the source of radiation (Brown & McDowell, 1962b). After irradiation, the surviving monkeys were used at Texas for several years in experiments on attention and learning (Sections II, E, and II, F). In the other study (Harlow *et al.*, 1956), four cynomolgus monkeys (*Macaca irus*) were tested on a variety of tasks before and after two of the monkeys were carried aloft by a balloon for exposure to cosmic radiation for 62 hours at altitudes above 90,000 feet. The rationale for the study lay in the inability to produce a low-energy heavy-nuclear radiation by accelerators then available, and the possibility that this type of radiation might have serious biological consequences. The dosage of radiation was unreported and had no apparent effect on appetite or on performance of learned behaviors.

A. Clinical Effects

Rhesus monkeys that receive near-lethal radiation (400 r) vomit within a few minutes and show obvious discomfort and lassitude. The next day, these animals respond efficiently when tested in the Wisconsin General Test Apparatus (WGTA; see Chapter 1 by Meyer *et al.* in Volume I). However, after 12 to 16 days, they are unable to resist infection as a consequence of declining numbers of white blood cells. They appear weak, have buccal swelling and runny noses. If they survive this period of crisis, they typically improve and may live many years. Very surprisingly, only 36 hours before death, monkeys that are debilitated to the point of nearly complete prostration will continue to respond efficiently to problems that offer rewards of food (see Section II, F, 1). Severely debilitated monkeys can also be motivated by electric shock (Melching & Kaplan, 1954), but the use of noxious stimulation instead of positive reinforcement is clearly unnecessary.

At the University of Washington, it was found that whole-body irradition in a single dose between 50 and 400 r is followed by anorexia, with 400 r producing the greatest effect (Fig. 1). Anorexia was most pronounced on the day of irradiation. However, some of the irradiated monkeys consumed the smallest amount on the next day, and even more monkeys showed greatest anorexia on a later day, especially the third day, in the first postexposure week. Two of five 400-r subjects, and most (but not all) subjects given lower doses, ate amounts of food in the normal range on one or more days in the first week, but a scattering of individually normal days does not establish a definite symptom-free period. Thus, the Washington findings contrast with a previously reported "symptom-free period following irradiation, in which almost all animals had undiminished appetites . . ." which ". . . lasted only two to three days in the 800 and 700 r groups and three to five days in the 600 and 400 r animals" (Eldred & Trowbridge, 1954, p. 68). Studies of

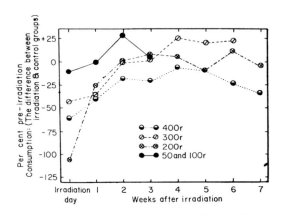

Fig. 1. Consumption of food by rhesus monkeys after irradiation. (Data for 300-r and 400-r groups from Ruch et al., 1962; data for other groups courtesy of R. W. Leary.)

irradiated humans seem to support the Washington findings. Hemplemann et al. (1952) pointed out that their severely injured Los Alamos patients, like a number of irradiated Japanese (Dunham et al., 1951), were definitely not well during the "latent" period. Therapeutic irradiation produces similar initial reactions of fatigue, nausea, vomiting, and anorexia; recovery may take as long as 5 days with no later recrudescence of symptoms.

B. General Behavior

In 1952, assisted by A. A. McDowell, J. P. Steele and I began studying the effects of radiation on the behavior of rhesus monkeys. A literature search did not provide information that would enable us to select appropriate measures of performance. Therefore, such measures were sought by observing the animals systematically. Standardized methods for reliably observing caged monkeys were lacking and had to be devised. Checklists were discarded because the investigators could not agree on the significant aspects of behavior to observe, and written transcriptions were discarded because they were unwieldy. The most satisfactory method was to observe without a screen and to speak normally into a microphone connected to a tape recorder. Subjects adapted to this very rapidly, and the observers could maintain constant surveillance. Speaking was in simple declarative sentences and continued constantly during each observation session in order to maintain an "observing set."

Sixteen young rhesus monkeys were caged in pairs. Each pair was observed continuously for 10 minutes once every 4 days. The monkeys were observed in this manner for 96 days in a preliminary study, then for 16 days immediately before and 32 days after 10 monkeys were irradiated with 400-r X-rays and the other 6 were sham irradiated (McDowell *et al.*, 1956). By sorting the observations, 74 categories of behavior were derived. Each category was classified either as rapid energy expenditure (bouncing, shaking the cage, pacing, or swinging from the roof of the cage), as visual survey (stationary monkey continuously shifting its visual regard), or as being oriented toward inanimate objects, toward the self, or toward another animal. Over-all activity decreased after irradiation. For this reason, the number of instances of each category of behavior within each observation period of 4 days was divided by the total number of times that any identifiable behavior occurred in that same 4-day period.

There was a reliable deficit in relative frequency of manipulating inanimate objects on postirradiation days 13–16 (Fig. 2), coincident with the most noticeable symptoms of sickness (see Section II, A). The num-

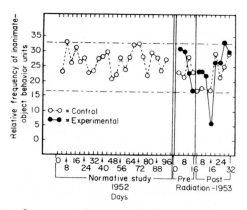

Fig. 2. Relative frequencies (times 100) of manipulation of inanimate objects by irradiated and sham-irradiated rhesus monkeys. Horizontal lines represent extremes of group performance in any 4-day period before irradiation. (After McDowell *et al.*, 1956.)

ber of minutes within which subjects shifted location showed a similar pattern of transient but marked deficit. Over the eight observation periods after irradiation, relative frequencies of aggression (Fig. 3)[2] and of rapid energy expenditure by the experimental subjects were reliably lower than before irradiation. On the other hand, the experimental animals showed relatively more self-grooming after irradiation than before. Similarly, Riopelle (1962, Table II) mentioned a decrease in social interactions by irradiated chimpanzees.

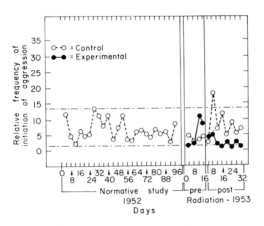

FIG. 3. Relative frequencies (times 100) of initiation of aggression by irradiated and sham-irradiated rhesus monkeys. (After McDowell et al., 1956.)

Hammack (1960) sought to determine the specificity of the radiation syndrome by observing the behavior of rhesus monkeys that had received doses of nitrogen mustard, a chemical agent that is biologically radiomimetic. The subjects were 24 rhesus monkeys divided equally into four groups. Nitrogen mustard, methyl bis (2-chloroethyl) amine hydrochloride, was administered intravenously in the following quantities per kilogram of body weight: group 1, 1.00 mg; group 2, 1.75 mg; group 3, 2.00 mg; and group 4, 3.00 mg. Control data were obtained from monkeys in group 2 prior to injection. Observations were recorded by depressing keys that operated clocks, counters, and an inkwriter. There were eight categories of behavior: shifts in location, cage manipulation, manipulation of wooden blocks placed in the cage, self-grooming, other-animal grooming, aggression, visual activity (fixed or shifting visual re-

[2]The anomalous increase in aggression by subjects in the control group during days 5–8 was probably due to the fact that the investigators broke the sleeping patterns of the animals, since later studies indicated that a similar increase could be induced by forced wakefulness (McDowell, 1954).

gard, not part of other behavior), and resting (no visual or other activity).

Relative frequencies of all categories of behavior changed significantly following treatment with nitrogen mustard. The effects were similar to those obtained by whole-body irradiation, but were maximal between days 5 and 9 rather than between days 13 and 18 as they were in the case of irradiation. The dose-response relations were complex and differed for relative frequency and relative duration of a particular category. Some measures of behavior were biphasic; e.g., a low dose of nitrogen mustard increased cage manipulation, whereas a high dose decreased it. Relative duration of self-grooming was related linearly to dose of nitrogen mustard, but relative frequency of self-grooming, although increasing with dosage, appeared to have an all-or-none character (Fig. 4).

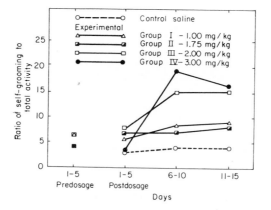

Fig. 4. Relative frequency (times 100) of self-grooming activities by nitrogen-mustard-treated and nontreated rhesus monkeys. (After Hammack, 1960.)

Leary and Ruch (1955) also reported changes in frequency of occurrence of selected responses following exposure to 400-r X-rays. On the day of irradiation, sitting increased sharply and there was a reliable decrease in chewing, in scratching and grooming, and in the number of movements between the front and the back of the cage. Each of these behaviors returned to its preradiation frequency within the first week or two. These brief changes thus began earlier than the transient decline in location shifts reported by McDowell *et al.* (1956). In the third and fourth weeks after irradiation, chewing became more frequent than it had been before irradiation. A similar trend was shown by scratching and grooming, a result similar to that for self-grooming in the South Dakota study.

Similar observations were also made at Texas, about a year after the last exposure to mixed γ-ray and neutron radiation (Davis et al., 1958a; McDowell, 1958). Rhesus monkeys were observed one at a time in a cage equipped with a rope and a cloth for manipulation and a mirror in which a monkey might view itself. The observer indicated which of 23 categories of behavior the monkey was engaged in at the end of each 15-second interval during 10-minute observation periods. The chronically irradiated monkeys made more cage-directed responses than the non-irradiated monkeys, but expended less energy and responded less often to sounds made by persons or monkeys outside the test room. On the basis of these observations and those by McDowell et al. (1956), it was hypothesized that irradiated and nonirradiated monkeys differ in distractibility or scope of attention. This hypothesis was later tested and confirmed (see Section II, E). Careful pre-experimental observation of general behavior was an efficient way to identify grossly which behaviors would reveal effects of irradiation in later experiments.

The surviving monkeys at South Dakota were recently observed by Davis (1962), using methods identical to those employed 7 years earlier by McDowell et al. (1956). There were marked changes in behavior of all animals. Particularly noticeable were increased self-oriented behavior and decreased social behavior. These changes are characteristic of aging in primates (see Chapter 12 by Riopelle and Rogers) and were observed in every subject, irradiated and nonirradiated. However, the changes were reliably less extensive in the irradiated monkeys than in the controls, thus contradicting the hypothesis that age changes in behavior are caused by gradually accumulating effects of radiation.

C. Manipulation and Motor Functions

Leary and Ruch (1955) measured strength by having rhesus monkeys pull a weighted tray to obtain food, and manipulation by the readings on a pedometer contained in a wood-and-plastic device that was placed in the cage. Strength of pull was not reliably impaired. However, radiation of 400 r depressed pedometer manipulation (Fig. 5) over a considerably longer time than suggested by the transient deficit in inanimate-object manipulation reported by McDowell et al. (1956).

Manipulation was further studied by Davis et al. (1956) using the six tests described below.

1. In a bent-wire detour problem, a candy Life Saver is threaded on a rigid wire having one or more 90° turns. The subject's task is to push the Life Saver along the path of the wire until it is free and can be eaten. As in detour problems which have been used to investigate "insightful" learning, and in contrast with such "blind" problems as the alley maze, the bent-wire test enables the subject to survey visually all relevant

aspects of the situation. Forty bent-wire problems were presented.
2. In a variant of the bent-wire task, punched paper poker chips were
used in place of candy Life Savers. Twelve difficult patterns were used.
3. The third task included four simple manipulation devices, one consisting of nine screw eyes that could be removed from holes in a board,
another of nine pairs of hooks and eyes, the third of nine hasps and
hasp staples. The fourth device was a composite of the elements in the
first three, and consisted of three units, each unit including a hasp restrained by a hook which in turn was restrained by a screw eye (see Fig.
1 in Chapter 13 by Butler). Puzzles were presented in both the living
cage and the WGTA, but manipulation was much more pronounced in
the former than the latter location.

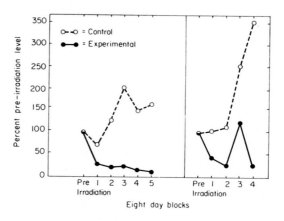

Fig. 5. Pedometer manipulation after a 400-r dose of radiation. Results of two
similar but separate experiments are shown. (After Leary & Ruch, 1955.)

4. A measure of oral manipulation of nonfood objects was suggested
by the observation of monkeys chewing on perches, puzzles, and cage
parts. A "sandwich" of two 3- by 3- by 1-inch pieces of pine board was
bolted together with a metal eyebolt and thrown to an animal in its
living cage. Each device was weighed nightly to determine the amount
of wood chewed from the blocks.
5. The fifth task was a six-unit mechanical puzzle that consisted of
common items of hardware arranged so that movement of parts of the
puzzle was possible only in a fixed order.
6. A patterned-strings problem consists of chains placed on a board in
a pattern with food attached to the end of one chain. The subject responds correctly by pulling the baited chain. Six difficult patterns were
used.

The six tests were presented to rhesus monkeys during 185 days of preliminary training, and then for 32 days immediately before the subjects were divided into irradiated and sham-irradiated groups and for 80 days after irradiation. Significant decrements in performance of the irradiated group on the first four tests occurred around the 16th postradiation day, but in every instance there was complete recovery by the 48th day. Individual monkeys sometimes stopped performing on particular puzzles, but no animal stopped performing entirely on all tests until death. The resulting syndrome appeared to be a loss in persistence rather than incoordination, as no tremors, paralysis, or disturbances of movement were noted.

One immediately wonders if loss in persistence is nothing more than generalized malaise. To determine if this was the case, Davis *et al.* (1958b) gave oral doses of 0.5 gm phenobarbitol sodium daily for 18 days to each of the animals that survived irradiation and to their controls. Sedation, like irradiation, affected performance on tasks requiring manipulation, but some of the differences resulting from sedation were opposite in direction to those produced by irradiation (radioantithetic), some in the same direction (radiomimetic), and still others unique to sedation. This indicated at least that there was not a common single syndrome of sickness for the two debilitating agents.

As indicated earlier, the irradiated survivors of these experiments received two additional doses of radiation. Performance on manipulation tasks was less affected by the second and third irradiations (Davis & Steele, 1963). Irradiated monkeys and their controls differed significantly after multiple doses of radiation on two tests of manipulation, the bent-wire detour problems and the patterned-strings tests. The speed of the irradiated monkeys in removing Life Savers clearly increased between the training that preceded the third irradiation and the training a year after (Fig. 6). Performance was facilitated on both easy patterns (1–8) and more difficult patterns (9–16). Increased speed was not an artifact of selective death of animals that were slow during original training. The irradiated animals solved fewer patterned-string problems than the controls, but this resulted, in part at least, from death of animals that solved a higher percentage of string problems during original training. Even more obvious selective effects were found for repeatedly-administered discrimination problems (see Section II, F, 2).

Two manipulation tasks were given to rhesus monkeys at Texas 3 through 5 months after the completion of irradiation (Davis *et al.*, 1958a). The first task, bent-wire detours, was described earlier in this section. The second, a finger-dexterity test, merely required monkeys to pick raisins out of small holes in a board, and the experimenter recorded the time taken to complete the task. Neither test differentiated between

Roger T. Davis

groups of irradiated and control monkeys. These results were similar to
the segment of data in Fig. 6 that was collected on bent-wire detour
problems during early chronic tests at South Dakota.

Nissen and several co-workers at the Yerkes Laboratories studied the
effects of radiation on the behavior of chimpanzees during 8 years following irradiation (Riopelle, 1962). Decreased frequency of response during

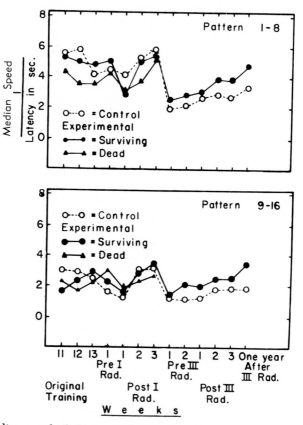

FIG. 6. Median speed of rhesus monkeys in obtaining candy on bent-wire detour
problems as a function of training and repeated irradiation. (After Davis & Steele,
1963.)

acute radiation sickness was reported for performance on bent-wire detours, poker-chip manipulation, and wood-block chewing, all of which
had shown transient decrements following irradiation of monkeys (Davis
et al., 1956). Irradiated chimpanzees also showed transient decrements
in rate of response in operant conditioning and in performance on the
elevator problem. The latter is a detour problem which allows the subject to survey visually the relevant aspects of the situation and requires

sustained and coordinated use of both hands (McDowell & Nissen, 1959). The irradiated chimpanzees showed no decrement in finger dexterity, strength of pull, or accuracy of performance on bent-wire detours and patterned-strings tests.

D. Food Preferences

There is, without doubt, a change in the preferences of irradiated monkeys for common laboratory foods. Leary (1955) reported that rhesus monkeys chose apple and carrot more frequently and peanut less frequently after whole-body irradiation than before. The changes were restricted to a period of 4 weeks after irradiation and were dependent on dose. It appeared to Leary that the change in preference was from higher to lower caloric value of the foods or possibly from dry to moist foods. Davis (1958) found that both irradiated and nonirradiated rhesus monkeys prefer foods in the order raisin, apple, potato, bread, and celery, but that the groups differ reliably in preferences for particular foods for as long as 18 months after irradiation. Both bread and raisins were more preferred and celery less preferred by irradiated than by nonirradiated monkeys. Although both studies showed that food preferences change after irradiation, Davis's results do not support Leary's interpretations, because bread was the driest food employed by Davis and raisins had the highest caloric value.

A very tempting explanation of changes in food preference, in light of the clearly-demonstrated conditioning of aversion for previously preferred liquids under irradiation (see Section I), is that during postirradiation anorexia monkeys eat only preferred foods and the taste of these foods is conditioned to nausea. However, Leary indicated that appetite and preference did not co-vary, and Davis showed that the most highly preferred food (raisins) was more highly preferred by the irradiated than by the nonirradiated subjects. A more plausible explanation is that the mouth, the throat, or the stomach is more sensitive to abrasion after irradiation and the abrasive foods, like peanuts in Leary's study and celery and raw potato in Davis's, are less preferred because they produce discomfort.

E. Attention and Distraction

As mentioned in Section II, B, the systematic observations of general behavior of rhesus monkeys at South Dakota and at Texas suggested that irradiated monkeys are less distractible or have a narrower scope of attention than nonirradiated monkeys. Harlow and Moon (1956) also suggested that irradiated monkeys may be less distractible because they are less active than normal.

The suggested change in scope of attention may be described as a change in the degree to which certain classes of stimuli influence behavior. Those stimuli that normally exert the greatest degree of control over behavior apparently control the behavior of irradiated monkeys still more strongly. In other words, if a particular stimulus is the one that a normal monkey is most likely to respond to in a certain situation, then an irradiated monkey in the same situation will have an even higher probability of response to that stimulus. This implies that other stimuli must be less likely to elicit responses in irradiated animals than in normal animals, or, in other words, that these other stimuli are less likely to distract an irradiated monkey than a nonirradiated monkey.

To explore the differences in responsiveness of the irradiated and nonirradiated rhesus monkeys at South Dakota, Grodsky (1954) tested the hypothesis that these two groups had different preferences for particular environmental stimuli. His apparatus was a large metal T-maze, the goal boxes of which provided different conditions of stimulation. There were eight conditions, paired in all possible combinations: (1) no manipulanda, (2) a wooden shelf, (3) wire strands, (4) a small window through which the subject could view toys, (5) shelf and wire strands, (6) shelf and window, (7) wire strands and window, and (8) shelf, wire strands, and window. The monkey entered the maze at the choice point from a transport cage which formed the stem, and the experimenter recorded the time that the monkey spent in each goal box during each 2-minute trial. Monkeys clearly preferred to remain in the box that afforded a view, as would be expected from previous studies of investigative behavior (see Chapter 13 by Butler). Irradiated monkeys were less affected by the presence or absence of a window than nonirradiated monkeys, but the difference barely missed statistical significance. However, the conclusion that irradiated monkeys are less responsive to stimuli outside the cage than are nonirradiated monkeys would seem warranted in the light of later research. For example, Harlow (1962) reported that irradiated monkeys opened the door of a Butler box less often than did nonirradiated monkeys. Frequency of opening the door decreased sharply in the week before death from accumulated doses of radiation.

McDowell and Brown (1963b) recently reported observations of the general behavior of male rhesus monkeys both with and without social distraction in the form of a female monkey 3 feet away from the subject's cage. The frequency of nondirected activities of nonirradiated males decreased markedly in the presence of the female. Males that had been irradiated as long as 6 years before the experiment showed smaller decreases in nondirected activities in the presence of the female, with the nondirected activities of high-dose groups showing practically no effect of the social stimulus.

That irradiation reduces the rhesus monkey's responsiveness to stimuli outside the cage was also confirmed experimentally using the WGTA (McDowell, 1958). Before each trial, a piece of grape was hidden in each of eight covered foodwells. The experimenter recorded how long the monkeys took to obtain all eight pieces under various conditions of distraction. Two experiments were performed, each using three conditions. The response times of both the irradiated and the nonirradiated monkeys were systematically influenced when the three conditions were no added stimuli, a hand organ playing outside the cage, and the cage door noisily opening to reveal mechanical puzzles. But when the conditions were no added stimuli, a floodlight flashing on the test tray, and a pounding noise, only the nonirradiated monkeys responded at reliably different rates under the three conditions. Thus, the irradiated subjects gave evidence of being distracted by stimuli within the cage but not by stimuli outside it.

Davis (1963) gave single-object and double-object delayed-response problems to rhesus monkeys at South Dakota 4 years after the third irradiation. Nonirradiated subjects had significantly longer latencies on double-object than on single-object problems, and the opposite trend (though not significant) characterized the irradiated subjects. The irradiated subjects apparently were not distracted by the presence of the second object, but the nonirradiated monkeys were. Independently, McDowell et al. (1963) obtained very similar results. Rhesus monkeys irradiated with high doses responded just as quickly on two-choice discrimination trials as they did to the positive object presented alone with position randomized, whereas low- and intermediate-dose groups were apparently distracted by the second object, and had longer latencies when it was present.

F. Learning and Transfer

1. IMMEDIATE EFFECTS

The initial investigations of learning in monkeys which received lethal or near lethal whole-body doses of radiation were conducted at Texas (e.g., Kaplan & Gentry, 1953, 1954; Kaplan et al., 1954; Rogers et al., 1954). In these exploratory studies, rhesus monkeys typically were trained on one particular discrimination task, irradiated, and retrained on on an equivalent problem. Using 108 discrimination-reversal problems, Warren et al. (1955) found that irradiated monkeys made more errors in the 10 postreversal trials than did controls. However, in the other studies the results were negative, and it became obvious that work with monkeys demanded an approach that was more conservative of animals and investigated a broader spectrum of behavior than was examined in the earlier work.

With the exception of Leary and Ruch (1955), all those who contemporaneously studied the effects of whole-body irradiation on the behavior of monkeys were particularly interested in effects on learning. Davis *et al.* (1956, 1958a), Harlow and Moon (1956), and Riopelle *et al.* (1956) reported the performance of irradiated and nonirradiated rhesus monkeys on a wide variety of learning tasks. The selection of tasks was specifically guided by sensitivity to particular types of brain damage. The tasks included avoidance learning, the Hamilton perseverance test (see Chapter 11 by Zimmermann and Torrey), simultaneous and successive object-quality-discrimination problems, pattern-discrimination problems (see Chapter 1 by Meyer *et al.* in Volume I), oddity and oddity-nonoddity problems (see Chapter 5 by French in Volume I), spatial and nonspatial delayed response (see Chapter 4 by Fletcher in Volume I), and the reduced-cue problem. This last task combines aspects of object discrimination and spatial delayed response in a two-trial problem. The first trial involves a nonspatial discrimination between two dissimilar objects. On the second trial, the distinguishing cues are removed, but food is hidden in the same position as on the first trial, enabling the subject to choose between two identical objects on the basis of position.

A few effects of radiation were reported, but for every reported effect on performance on a given task there were one or two reports of no effect on the same or similar tasks. Dissimilar experience of irradiated and nonirradiated subjects, selection by death of good or poor performers, and the general statistical vagaries of small samples of monkeys and large samples of tasks could easily account for the discrepant results. But there were very few exceptions to account for: in general, performance of learned behaviors was not affected by irradiation, and this was true whether the behaviors were learned before or after irradiation or during the intervals between repeated doses of radiation.

Undoubtedly the most startling result was that well-trained monkeys did not have to consume food for it to be rewarding. Irradiated monkeys maintained their accuracy of choice in spite of debilitation that approached terminal coma (Davis *et al.*, 1956; Harlow & Moon, 1956). But Davis (1963) found that this result was not specific to irradiated subjects; well-trained rhesus monkeys, whether irradiated or not, continued to perform accurately on delayed-response problems when a piece of celery was substituted for a raisin during the delay (cf. Tinklepaugh, 1928). In fact, even substituting spitballs, or completely eliminating reward on 24 consecutive trials, did not increase the number of errors.

2. DELAYED EFFECTS

Quite unlike the pattern of negative data that characterized the studies of the effects of irradiation within the first year are the often-significant chronic consequences of irradiation. But caution is needed in interpreting these delayed effects because, without fail, deaths occurred among the original groups of subjects. Thus, when the performances of irradiated animals and their controls differ, there is always the question of whether the differences are progressive consequences of irradiation or are due to selection of brighter or duller subjects through death.

Davis and Steele (1963) reanalyzed data collected at South Dakota over a 7-year span and compared the scores of the irradiated animals that ultimately survived and of those that died. The differences were remarkable and involved both discrimination and oddity problems, neither of which was affected immediately by radiation (Fig. 7). It is clear

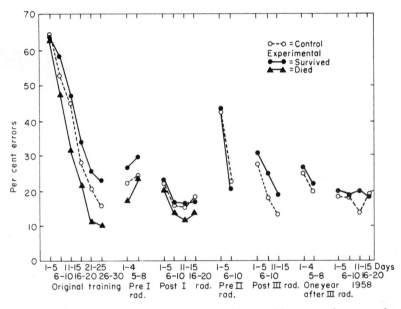

FIG. 7. Performance of rhesus monkeys on oddity problems as a function of training and repeated irradiation. (After Davis & Steele, 1963.)

that the superior animals died leaving survivors that were significantly inferior to the controls. Performance on delayed-response and reduced-cue problems revealed both selective and progressive changes. Selection could very well underlie other delayed effects summarized in this section.

Many of the chronic effects that have been reported were discovered at Texas, using the large groups of laboratory- and field-irradiated rhesus monkeys described by Brown and McDowell (1962a). These investigators account for many of their findings in terms of "decreased distractibility or, in other words, increased concentration of attention. The general resultant for learning is facilitation of performance on tasks placing a premium on attention to the site of food reward and decrement in performance on tasks requiring attention to peripherally placed stimuli" (Brown & McDowell, 1962a, p. 745).

The tasks on which the performance of irradiated monkeys has been most clearly shown to be facilitated or impaired are those tasks that most clearly fit Brown and McDowell's descriptions. Thus, irradiated rhesus monkeys clearly perform better than nonirradiated controls on spatial delayed response (McDowell & Brown, 1959b) and on the closely related reduced-cue problem (McDowell & Brown, 1958), tasks that undoubtedly require attention to the locus of food.

Conversely, two types of task requiring attention to stimuli placed at some distance from the sites of food (and of responses–cf. Chapter 1 by Meyer *et al.* in Volume I) have revealed better performance by nonirradiated monkeys than by irradiated subjects. One of these tasks (Brown *et al.*, 1959) involved two discrimination problems each day, presented in the usual way with a pair of stimulus objects covering the foodwells in a WGTA. During training on the first problem each day, the objects for the second problem were placed 6 inches behind the objects that covered the foodwells. The left-right positions of the objects of both pairs varied together, so that one object of each pair was consistently on the same side as food. The positive object in the second problem was the object that had been placed 6 inches behind the object covering food in the first problem. The nonirradiated monkeys showed positive transfer to the second problem, whereas the irradiated monkeys did not.

Because of the intimate relation between motivation and attention (cf. Birch, 1945), Davis and Lovelace (1963) repeated this experiment using nonpreferred (celery) as well as preferred (raisin) rewards. With raisins, but not with celery, the results corroborated the finding of Brown *et al.* that irradiated subjects ignore auxiliary discriminanda. Thus, attention to cues depends on effectiveness of rewards. Changes in reward effectiveness may be sufficient to explain changes in attention to cues following irradiation of rhesus monkeys.

In the second type of task involving spatial separation of cues from responses and rewards, the response and reward loci were also separated from each other. Using such an arrangement, McDowell and Brown (1963a) recently trained rhesus monkeys on a simple discrimination between a red and a gray block of wood. On each trial, the objects were placed at the rear of the test tray, one almost 8 inches to the left, and the other the same distance to the right, of a centered opaque barrier. Two chains ran back from the front of the tray, past either side of the barrier, ending behind the barrier out of view of the monkey. A raisin or a piece of apple was impaled on a nail attached to the end of the chain that was closer to the positive object. The results (Fig. 8) showed a pronounced interaction between the variables of sex and irradiation. The irradiated male subjects improved more rapidly with training than the nonirradiated males, but the irradiated females improved less rapidly than the nonirradiated females. McDowell and Brown interpreted their findings in terms of previously demonstrated sex differences in attending by monkeys (McDowell *et al.*, 1960). Female monkeys, which normally have a narrower scope of attention than males, may have suffered so

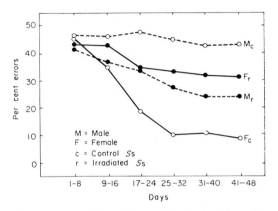

FIG. 8. Performance of male and female irradiated and nonirradiated rhesus monkeys on a discrimination problem with stimulus objects spatially separated from response loci. (After McDowell & Brown, 1963a.)

much reduction in scope of attention after irradiation that relevant cues were not used and performance worsened. Since normal male monkeys are more distractible, the reduction in scope of attention following irradiation may have reduced the number of distracting stimuli, thus improving performance. Although McDowell and Brown parsimoniously regarded the results as reflecting an irradiation-induced narrowing of attention for both sexes, they implicitly assumed that differences in scope of attention are additive.

With male rhesus monkeys as subjects, McDowell (1960) used the same chain-and-barrier arrangement to test transfer of a discrimination that was learned with the stimulus objects placed over foodwells in the usual way. Irradiated and nonirradiated monkeys made equal numbers of errors in original learning, but when the stimulus objects were placed at the rear of the tray the various groups made different numbers of errors. The effect was nonmonotonic: the monkeys that had received the highest dose of radiation performed more poorly than the nonirradiated controls, but the low-dose group performed somewhat better than the controls, and the intermediate-dose group performed best. A possible interpretation of this result is similar to McDowell and Brown's (1963a) interpretation of the interaction of sex and irradiation. Low and intermediate doses of radiation may have reduced the distractibility of the male subjects and thus improved transfer, while a high dose may have narrowed the scope of attention still further, partially excluding even the peripherally-placed relevant cues. Such an interpretation is supported by Brown and McDowell's (1962a, Fig. 14) more detailed presentation of the interaction of sex and irradiation summarized in Fig. 8. The number of errors made by female subjects increased monotonically as radiation dose increased, but the male subjects' performance was related to dose in the same nonmonotonic fashion as in McDowell's experiment.

Another type of task that has revealed deficits of irradiated rhesus monkeys is visual pattern discrimination. In several experiments, discrimination of a broken circle from a whole circle has been used as a test of visual acuity. For training, a circle with a large gap is paired with the whole circle; then the discrimination is transferred to progressively smaller gaps. Five to 6 months after irradiation, the monkeys at Texas that had received the highest dose made more errors on the transfer problems than did nonirradiated monkeys (Davis et al., 1958a). Three years after exposure, the same animals continued to show this deficit (Brown & McDowell, 1960). The intermediate-dose group, which previously had not differed from the control group, also made more errors at that time. Brown and McDowell pointed out that proficiency on the first and easiest discrimination was directly related to radiation dose, so the irradiated monkeys could not have been deficient in basic learning ability. But the deficiency may not have been one of visual acuity, either. First, performance differences appeared quite early in the series of transfer problems, although even the last and most difficult problem was solved to a rigid criterion by all subjects. Second, using similar stimulus patterns but modified procedures, Lovelace and Davis (1963) tested the monkeys at South Dakota 4.5 years after the third irradiation and found no significant differences in performance of irradiated and nonirradiated subjects. Their transfer-testing procedure apparently minimized errors associated with sudden introduction of a more difficult problem. Greater proneness to such errors, rather than poorer vision, may characterize the irradiated primate.

Impaired pattern-discrimination performance, whatever its cause, has been found in at least two other studies of irradiated primates. Riopelle (1962) reported that, several years after irradiation, chimpanzees were inferior to control subjects in learning to discriminate fine visual detail and in accuracy of four-choice oddity performance on problems requiring discrimination of wallpaper patterns. A similar finding was reported by Heimstra *et al.* (1957), who gave 300 five-trial "random pattern" discrimination problems (see Chapter 1 by Meyer *et al.* in Volume I) to rhesus monkeys that survived two 400-r doses of radiation at South Dakota. The performance of the irradiated monkeys was inferior to that of identically-trained nonirradiated controls when the subjects were given 5 problems each day. However, when 25 problems were given each day, the nonirradiated subjects performed more poorly than they did on 5 problems per day. This effect, which reflected a deterioration of performance from problem to problem within 25-problem sessions, was not shown by the irradiated subjects. Such a motivational difference between irradiated and nonirradiated monkeys may result from the irradiated animals' increased preference for the preferred foods that were used as rewards (see Section II, D).

One type of pattern-discrimination problem has shown facilitation, rather than impairment, of performance of irradiated monkeys (McDowell & Brown, 1962). About 2 years after exposure to nuclear radiations at the Nevada Test Site, rhesus monkeys were trained to discriminate the presence or absence of a 1-inch black dot in the center of a white card. When the diameter of the dot was repeatedly halved to $\frac{1}{32}$ inch as a test of visual acuity, proficiency on the more difficult discriminations was directly related to radiation dosage. McDowell and Brown attributed the results to the increased attentiveness of irradiated monkeys to the site of food reward.

A similar explanation was advanced by McDowell and Brown (1963d) for their results in an experiment on extinction. After 24 trials of training on an object-discrimination problem, the same objects were presented for 24 more trials with no food under either object. During these extinction trials, responses with latencies of 3 seconds or more were made earlier and more frequently by irradiated than by nonirradiated rhesus monkeys, perhaps because irradiation facilitated attention to the food or nonfood outcomes of responses.

Faster extinction might suggest facilitation of discrimination-reversal learning, but apparently the direction of irradiated monkeys' responses is more stable than the latency. During reversal training on a two-choice object-quality-discrimination problem, the number of errors before the first correct response was directly related to radiation dose (McDowell & Brown, 1963e). This result accounts for the poor performance of irradiated monkeys on a reversal-learning-set procedure involving only 10 trials per reversal (see Section II, F, 1). But this perseveration of response to the previously positive object does not last long enough to significantly affect the total number of trials or errors when the reversal is learned to a high criterion (McDowell & Brown, 1959a, 1963e). This suggests that once an irradiated monkey has first responded correctly in reversal learning, it may respond more consistently to the new positive object than does a nonirradiated monkey. Greater consistency of response could account for the finding that negative transfer, as measured by the ratio of postreversal to prereversal errors, is decreased by irradiation (McDowell & Brown, 1959a). Perseveration of response to a previously positive object and consistency of response to a new positive object both suggest that once an irradiated monkey has learned to obtain food by responding to a particular object, its attention is focused on that object.

Attentiveness to a particular positive object is also suggested by the results of an experiment by Overall *et al.* (1960). Rhesus monkeys were trained on intermediate-size discrimination and then tested for transposition. Failure to transpose was directly related to radiation dose, suggesting that as the dose increased, the animals attended more to the absolute size of the positive object and less to its intermediate-size relation to the negative objects. It should be mentioned, however, that irradiated chimpanzees showed no decrement on an unspecified test of size transposition (Riopelle, 1962).

Several other studies have found differences in response by irradiated and nonirradiated rhesus monkeys when stimulus objects are changed or added during discrimination training. In one such study, Overall and Brown (1958) interpolated trials on which a pair of identical black (positive) objects appeared during presentation of a black-white discrimination problem. Nonirradiated monkeys and those irradiated with low doses responded to the position that had been occupied by the positive object on the previous trial, but monkeys that had received high doses did not. Attention to the positive object may have excluded attention to its position, which was irrelevant to obtaining food.

Two object-quality-discrimination experiments at South Dakota involved the introduction of new objects which differed from each other and from the objects used on conventional discrimination trials. In the first study (Davis, 1961), three additional negative objects were presented on the even-numbered trials of ordinary 10-trial problems. Performance of irradiated and nonirradiated monkeys did not differ on odd-numbered (two-choice) trials, but performance of irradiated monkeys was more seriously impaired than that of the controls on even-numbered (five-choice) trials. In the second study (Fitzgerald & Davis, 1960), each two-choice problem was first presented for one, three, or five conventional trials. Then a trial was given in which either the positive or the negative object was replaced by three new objects. All subjects continued to choose the positive object when the negative object was eliminated. When the positive object was eliminated, nonirradiated monkeys responded to the original negative object more often than did the irradiated monkeys. This difference also occurred when errors were made on five-choice trials in the first study.

III. EFFECTS OF HIGH-DOSE IRRADIATION OF THE BRAIN

Doses of radiation above 2,500 r produce histological changes in the brain. Such doses are above the LD-100 for whole-body irradiation, so they must be administered focally if behavior is to be studied for very long after irradiation. High doses are administered in two ways. One is to insert needles of a radioactive material into the subject's brain; the other is to expose part or all of the brain through holes in lead shields. The latter method may or may not involve exposing the brain surgically, and requires special care to shield the eyes and jaws to avoid corneal and retinal damage and buccal infections. Three major current programs (at the Universities of Nebraska, Texas, and Wisconsin) have been concerned with high-dose irradiation of the brain or head, and the last two programs have used rhesus monkeys as subjects.

The work at Texas (Davis & McDowell, 1962) limited X-radiations to selected regions of the head by using lead caps. The prefrontal cortex was exposed in one group, the inferior parietal lobule and temporal lobe in another, and both the anterior and posterior foci in a third group. Sham-irradiated monkeys served as a control group. The animals were given preliminary training in a WGTA on a series of tasks which included object-quality-discrimination problems, bent-wire detour problems, patterned-strings tests, the elevator problem, and food-preference tests. Each of these tasks was selected because it was diagnostic either of brain damage or of whole-body irradiation. In addition, subjects were observed systematically, one at a time, in a specially-constructed observation cage. During 12 days before irradiation the tasks were presented in a massed fashion to obtain a daily measure of performance on every task.

Each subject in the three experimental groups was given 3,000 r to the head after its baseline performance had been determined on each task. Training was resumed for 27 days. Then a second dose of 3,000 r was administered in the same focus as before, and a 30-day period of intensive training followed. In addition, the performance of the animals that survived was studied during the following 6 years.

The results for the first part of the study clearly indicated that X-irradiation of parts of the heads of monkeys, in two 3,000-r doses spaced nearly a month apart, did not have the same effect as surgical damage to comparable parts of the brain. The only significant change in performance in the WGTA was a decrement on the elevator task by subjects irradiated both anteriorly and posteriorly. Other significant changes in behavior were noted in the observation cage. Rapid energy expenditure (see Section II, B) increased in animals whose frontal lobes had been irradiated. This change was similar to the forced pacing of animals with frontal-lobe lesions. There was also an increase in visual survey by animals that had been irradiated through both the anterior and posterior ports. Irradiated subjects maintained their high pre-exposure levels of performance throughout the first 30 days after the second irradiation unless death intervened, and in every case death was preceded by marked and sudden onset of neurological symptoms, especially trunk ataxia. Every monkey that had been irradiated both anteriorly and posteriorly died within 210 days of the second irradiation.

The chronic survivors of this experiment ultimately showed effects that did indeed resemble those of surgical brain damage, but not those of whole-body irradiation. About 2 years after irradiation, the survivors were trained to discriminate broken circles from whole circles (McDowell & Brown, 1960). The animals surviving irradiation of the posterior association cortex made more errors than nonirradiated monkeys on the more difficult discriminations, which McDowell and Brown attributed to deficient visual acuity. Animals with prior frontal-lobe irradiation, however, were inferior to the other groups throughout training, suggesting a basic deficit in the ability to learn the discrimination. The frontally-irradiated monkeys also took longer to reach criterion on a three-choice discrimination-reversal problem than did nonirradiated controls or animals irradiated posteriorly (McDowell *et al.*, 1961). Unlike animals with whole-body irradiation, those with head irradiation were similar to nonirradiated controls in transferring a learned discrimination when the stimulus objects were placed at a distance from the chains to which responses were made (McDowell & Brown, 1959c). Surprisingly, the delayed-response test, ordinarily very sensitive to frontal-lobe damage, did not distinguish irradiated and nonirradiated subjects (McDowell *et al.*, 1961).

When another monkey was placed in a cage 3 feet away (cf. Section II, E), posteriorly-irradiated male monkeys made fewer nondirected and more cage-directed responses, and their social responsiveness was unaffected by the sexual status of the stimulus monkey (McDowell & Brown, 1963c). Under the same conditions, the frequency of nondirected responses by nonirradiated monkeys remained constant, but these monkeys made fewer responses to cage parts and were more responsive to receptive females than to other stimulus monkeys.

The work at Wisconsin (Harlow, 1958) involved the technique, described earlier by Settlage and Bogumill (1955), of implanting cobalt-

60 needles into the cerebral cortex of monkeys in order to produce small but prominent lesions. Doses were computed as the total number of roentgens delivered 5 mm from the center of the source. Bilateral implantation of needles that gave 8,000-r doses and unilateral placement of a needle giving 4,000 r had the same effect: either operation sharply depressed performance on tests of previously learned behaviors, including delayed response, two-trial discrimination, and patterned strings. Performance improved steadily from chance level to the high preoperative levels, complete recovery taking about 20 days. Wide variations in the placement of the needles produced remarkably similar symptoms. If subjects were irradiated, and an area of 5- to 6-mm radius surrounding the lesion was removed surgically, the effect remained. However, comparable surgical lesions in nonirradiated animals did not produce the transitory deficit common to irradiated subjects, nor did radiation doses of 3,000 r or less. Harlow posited a diaschisis rather than necrosis, tissue loss, or motivational change to explain these findings. He described diaschisis as the transmission through association or commissural pathways of atypical patterns of neuron discharge producing "reverberations at a distance" which resulted in behavioral decrement. Harlow presented evidence that the fibers of the corpus callosum played an important role in producing the transient but very striking cortical irradiation syndrome.

IV. SUMMARY AND INTERPRETATION

In adult monkeys, the clearest consequences of whole-body irradiation in the range of LD-50 are: a reduced field of attention, lowered distractibility, intensified preference for foods that are typically preferred (and soft in texture) and lessened preference for foods that are infrequently chosen (and coarse in texture). Immediately after irradiation there is a loss in appetite, less manipulation, less rapid expenditure of energy, and less initiation of aggression. Appetite and manipulation return to normal quickly, but the other responses remain depressed indefinitely and maybe permanently. Effects that appear long after irradiation can be due either to selection through death or to progressive consequences of irradiation.

High doses of radiation have been given directly to the brain with cobalt-60 needles and with X-ray beams. Gamma radiation from cobalt 60 produces a syndrome of short duration but striking extent, and the tasks affected are those that have been diagnostic of surgical brain damage. However, because behavior returns to normal, these effects seem to be of stimulation rather than damage. In contrast, X-radiation produces no immediate effects on tasks that are diagnostic of surgical brain damage. Subjects may die after sudden onset of ataxia, but if they survive they manifest a chronic syndrome that resembles some effects of brain damage.

Within the last few years, investigators have progressed from studying the effects of irradiation on maze learning by rats or delayed response by monkeys to rather thorough exploration of several behaviors that respond to radiation. This progression has been part of and may have even contributed to a *Zeitgeist* in animal psychology that admits topics other than learning and that accepts relatively atheoretical work. However, the time has come to try a modest bit of theorizing to aid in summarizing the data. The most parsimonious interpretation must combine the various attentional and motivational components of the syndrome. It must also take into account the generally accepted notion that the brain is not damaged by levels of whole-body radiation that produce discernible effects, yet also include the fact that the effects persist.

In summarizing a group of papers on the effects of ionizing radiation on behavior, Lindsley pointed out that some of the symptoms of monkeys that are irradiated to the head or with whole-body doses suggest involvement of the reticular formation (Haley & Snider, 1962, pp. 751–752). However, the evidence that has been presented in this chapter indicates that a greater change in attention and motivation is caused by whole-body irradiation than by irradiation of the head in 3,000-r doses, and that neither type of irradiation caused immediate intellectual loss. Since the attentional and motivational effects are chronic, a stimulating effect of irradiation can be discounted. However, irradiation could chronically and indirectly stimulate the nervous system.

The tissue of the gastrointestinal tract is very sensitive to radiation; 5 to 10 r causes local inflammation of the intestines. Furthermore, there is an ample supply of receptors in this tissue for mediating pain. Finally, chronic abdominal pain can arise from intestinal dysfunction (Grimson *et al.*, 1947). It seems reasonable that permanent changes in the intestines could provide a sensory pattern that persisted and bombarded the reticular formation. This in turn could have permanent effects on motivation and attention through the mechanism discovered by Hernández-Peón and Scherrer (1955). These investigators showed that stimulating one sensory modality inhibits afferent transmission in another, and they suggested that ". . . this selective inhibitory mechanism plays an important role during 'focusing of attention' for sensory perception" (Hernández-Peón & Scherrer, 1955, p. 71). A chronic discharge from small irritative lesions of the gut could easily inhibit other sensory inputs to the reticular formation. The explanation is congruous with the findings that radiation, especially to the body, is arousing (Hunt & Kimeldorf, 1962) and useful as an unconditioned stimulus for avoidance (Garcia *et al.*, 1961). Although Hunt and Kimeldorf suggested that arousal may result from direct stimulation of the nervous system, it seems more plausible that receptors in the vicinity of cells that are damaged or destroyed would be stimulated and continue to be stimulated chronically.

REFERENCES

Birch, H. G. (1945). Role of motivational factors in insightful problem solving. *J. comp. Psychol.* **38**, 295.

Brown, W. L., & McDowell, A. A. (1960). Visual acuity performance of normal and chronic irradiated monkeys. *J. genet. Psychol.* **96**, 133.

Brown, W. L., & McDowell, A. A. (1962a). Some effects of radiation on psychologic processes in rhesus monkeys. *In* "Response of the Nervous System to Ionizing Radiation" (T. J. Haley & R. S. Snider, eds.), pp. 729–746. Academic Press, New York.

Brown, W. L., & McDowell, A. A. (1962b). Behavioral changes as a function of ionizing radiations. *In* "Effects of Ionizing Radiation on the Nervous System." International Atomic Energy Agency, Vienna.

Brown, W. L., Carr, R. M., & Overall, J. E. (1959). The effect of chronic whole-body irradiation on peripheral cue associations. *J. gen. Psychol.* **61**, 113.

Claus, W. D. (1958). "Radiation Biology and Medicine." Addison-Wesley, Reading, Massachusetts.

Davis, R. T. (1958). Latent changes in the food preferences of irradiated monkeys. *J. genet. Psychol.* **92**, 53.

Davis, R. T. (1961). Discriminative performance of monkeys irradiated with X-rays. *Amer. J. Psychol.* **74**, 86.

Davis, R. T. (1962). Supplementary report: Effects of age and radiation on gross behavior of monkeys. *Psychol. Rep.* **11**, 738.

Davis, R. T. (1963). Chronic effects of ionizing radiations and the hypothesis that irradiation produces aging-like changes in behavior. *J. genet. Psychol.* **102**, 311.

Davis, R. T., & Lovelace, W. E. (1963). Variable rewards and peripheral cues in discriminations by irradiated and nonirradiated monkeys. *J. genet. Psychol.* **103**, 201.

Davis, R. T., & McDowell, A. A. (1962). Performance of monkeys before and after irradiation to the head with X-rays. *In* "Response of the Nervous System to Ionizing Radiation" (T. J. Haley & R. S. Snider, eds.), pp. 705–718. Academic Press, New York.

Davis, R. T., & Steele, J. P. (1963). Performance selection through radiation death in rhesus monkeys. *J. Psychol.* **56**, 119.

Davis, R. T., McDowell, A. A., Deter, C. W., & Steele, J. P. (1956). Performance of rhesus monkeys on selected laboratory tasks presented before and after a large single dose of whole-body X radiation. *J. comp. physiol. Psychol.* **49**, 20.

Davis, R. T., Elam, C. B., & McDowell, A. A. (1958a). Latent effects of chronic whole-body irradiation of monkeys with mixed source radiation. USAF Sch. Aviat. Med., Rep. No. 57–59.

Davis, R. T., McDowell, A. A., Grodsky, M. A., & Steele, J. P. (1958b). The performance of X-ray irradiated and nonirradiated rhesus monkeys before, during, and following chronic barbiturate sedation. *J. genet. Psychol.* **93**, 37.

Dunham, C. L., Cronkite, E. P., LeRoy, G. V., & Warren, S. (1951). Atomic bomb injury: radiation. *J. Amer. med. Ass.* **147**, 50.

Eldred, E., & Trowbridge, W. V. (1954). Radiation sickness in the monkey. *Radiology* **62**, 65.

Fitzgerald, R. D., & Davis, R. T. (1960). The role of preference and reward in the selection of discriminanda by naive and sophisticated rhesus monkeys. *J. genet. Psychol.* **97**, 227.

Furchtgott, E. (1956). Behavioral effects of ionizing radiations. *Psychol. Bull.* **53**, 321.

Furchtgott, E. (1963). Behavioral effects of ionizing radiations: 1955–61. *Psychol. Bull.* **60**, 157.

Garcia, J., Kimeldorf, D. J., & Hunt, E. L. (1961). The use of ionizing radiation as a motivating stimulus. *Psychol. Rev.* **68**, 383.

Grimson, K. S., Hesser, K. F., & Kitchen, W. W. (1947). Early clinical results of transabdominal celiac and superior mesenteric ganglionectomy, vagotomy, or trans-thoracic splanchnicectomy in patients with chronic abdominal visceral pain. *Surgery* **22**, 230.

Grodsky, M. A. (1954). Preference scales of monkeys using visual and auditory stimuli. Unpublished Master's thesis, University of South Dakota.

Haley, T. J., & Snider, R. S. (eds.) (1962). "Response of the Nervous System to Ionizing Radiation." Academic Press, New York.

Hammack, B. A. (1960). Changes in free cage behavior of rhesus monkeys as a function of differential dosages of nitrogen mustard. *J. genet. Psychol.* **96**, 275.

Harlow, H. F. (1958). Behavioral contributions to interdisciplinary research. *In* "Biological and Biochemical Bases of Behavior" (H. F. Harlow & C. N. Woolsey, eds.), pp. 3–23. Univer. Wisconsin Press, Madison.

Harlow, H. F. (1962). Effects of radiation on the central nervous system and on behavior—general survey. *In* "Response of the Nervous System to Ionizing Radiation" (T. J. Haley & R. S. Snider, eds.), pp. 627–644. Academic Press, New York.

Harlow, H. F., & Moon, L. E. (1956). The effects of repeated doses of total-body X radiation on motivation and learning in rhesus monkeys. *J. comp. physiol. Psychol.* **49**, 60.

Harlow, H. F., Schrier, A. M., & Simons, D. G. (1956). Exposure of primates to cosmic radiation above 90,000 feet. *J. comp. physiol. Psychol.* **49**, 195.

Heimstra, N. W., Davis, R. T., & Steele, J. P. (1957). Effects of various food deprivation schedules on the discrimination learning performance of monkeys irradiated with X-ray irradiation. *J. Psychol.* **44**, 271.

Hemplemann, L. H., Lisco, H., & Hoffman, J. G. (1952). The acute radiation syndrome: a study of nine cases and a review of the problem. *Ann. intern. Med.* **36**, 279.

Hernández-Peón, R., & Scherrer, H. (1955). "Habituation" to acoustic stimuli in cochlear nucleus. *Fed. Proc.* **14**, 71. (Abstract.)

Hollaender, A. (ed.) (1954–1956). "Radiation Biology," 3 vols. McGraw-Hill, New York.

Hunt, E. L., & Kimeldorf, D. J. (1962). Evidence for direct stimulation of the mammalian nervous system with ionizing radiation. *Science* **137**, 857.

Kaplan, S. J., & Gentry, G. (1953). The effect of sublethal dose of X-irradiation upon transfer of training in monkeys. USAF Sch. Aviat. Med., Proj. No. 21–3501–0003 (Rep. No. 4).

Kaplan, S. J., & Gentry, G. (1954). Some effects of a lethal dose of X-radiation upon memory: A case history study. USAF Sch. Aviat. Med., Proj. No. 21–3501–0003 (Rep. No. 2).

Kaplan, S. J., Gentry, G., Melching, W. H., & Delit, M. (1954). Some effects of a lethal dose of X-radiation upon retention in monkeys. USAF Sch. Aviat. Med., Proj. No. 21–3501–0003 (Rep. No. 8).

Lea, D. E. (1955). "Actions of Radiations on Living Cells," 2nd ed. Cambridge Univer. Press, London and New York.

Leary, R. W. (1955). Food-preference changes of monkeys subjected to low-level irradiation. *J. comp. physiol. Psychol.* **48**, 343.

Leary, R. W., & Ruch, T. C. (1955). Activity, manipulation drive, and strength in monkeys subjected to low-level irradiation. *J. comp. physiol. Psychol.* **48**, 336.

Lindgren, M. (1958). On tolerance of brain tissue and sensitivity of brain tumors to irradiation. *Acta Radiol. Suppl.* **70**, 1.

Lovelace, W. E. & Davis, R. T. (1963). Minimum-separable visual acuity of rhesus monkeys as a function of aging and whole-body radiation with X-rays. *J. genet. Psychol.* **103**, 251.

McDowell, A. A. (1954). The immediate effects of single dose whole body X radiation upon the social behavior and self-care of caged rhesus monkeys. *Amer. Psychologist* **9**, 423. (Abstract.)

McDowell, A. A. (1958). Comparisons of distractibility in irradiated and nonirradiated monkeys. *J. genet. Psychol.* **93**, 63.

McDowell, A. A. (1960). Transfer by normal and chronic whole-body irradiated monkeys of a single learned discrimination along a peripheral cue gradient. *J. genet. Psychol.* **97**, 41.

McDowell, A. A., & Brown, W. L. (1958). Facilitative effects of irradiation on performance of monkeys on discrimination problems with reduced stimulus cues. *J. genet. Psychol.* **93**, 73.

McDowell, A. A., & Brown, W. L. (1959a). A comparison of normal and irradiated monkeys on an oddity-reversal problem. *J. genet. Psychol.* **95**, 105.

McDowell, A. A., & Brown, W. L. (1959b). Latent effects of chronic whole-body irradiation upon the performance of monkeys on the spatial delayed-response problem. *J. gen. Psychol.* **61**, 61.

McDowell, A. A., & Brown, W. L. (1959c). Transfer by normal and chronic focal-head irradiated monkeys of a single learned discrimination along a peripheral cue gradient. *USAF Sch. Aviat. Med.*, Rep. No. 59–18.

McDowell, A. A., & Brown, W. L. (1960). Visual acuity performance of normal and chronic focal-head irradiated monkeys. *J. genet. Psychol.* **96**, 139.

McDowell, A. A., & Brown, W. L. (1962). Sex and radiation as factors in learning performance by rhesus monkeys on a series of dot discrimination problems. *J. genet. Psychol.* **101**, 273.

McDowell, A. A., & Brown, W. L. (1963a). Sex and radiation as factors in peripheral cue discrimination learning. *J. genet. Psychol.* **102**, 261.

McDowell, A. A., & Brown, W. L. (1963b). Social distractibility of normal and previously irradiated male rhesus monkeys. *J. genet. Psychol.* **103**, 309.

McDowell, A. A., & Brown, W. L. (1963c). Some evidence of "psychic blindness" in monkeys with focal-head irradiation of the temporal lobes. *J. genet. Psychol.* **103**, 317.

McDowell, A. A., & Brown, W. L. (1963d). Some differences in extinction by normal and whole-body irradiated monkeys. *J. genet. Psychol.* **103**, 335.

McDowell, A. A., & Brown, W. L. (1963e). Some comparisons of normal and irradiated monkeys on between-day reversal learning. *J. genet. Psychol.* **103**, 369.

McDowell, A. A., & Nissen, H. W. (1959). Solution of a bi-manual coordination problem by monkeys and chimpanzees. *J. genet. Psychol.* **94**, 35.

McDowell, A. A., Davis, R. T., & Steele, J. P. (1956). Application of systematic direct observational methods to analysis of the radiation syndrome in monkeys. *Percept. mot. Skills* **6**, 117.

McDowell, A. A., Brown, W. L., & McTee, A. A. (1960). Sex as a factor in spatial delayed-response performance by rhesus monkeys. *J. comp. physiol. Psychol.* **53**, 429.

McDowell, A. A., Brown, W. L., & White, R. K. (1961). Oddity-reversal and delayed-response performance of monkeys previously exposed to focal-head irradiation. *J. genet. Psychol.* **99**, 75.

McDowell, A. A., Brown, W. L., & Wicker, J. E. (1963). Effects of radiation exposure on response latencies of rhesus monkeys. *J. genet. Psychol.* **102**, 225.

Melching, W. H., & Kaplan, S. J. (1954). Some effects of a lethal dose of X-radiation upon retention: studies of shock avoidance motivation. USAF Sch. Aviat. Med., Proj. No. 21–3501–3003 (Rep. No. 9).

Overall, J. E., & Brown, W. L. (1958). Narrowing of attention as a chronic effect of sublethal radiation. USAF Sch. Aviat. Med. Rep. No. 58–27.

Overall, J. E., Brown, W. L., & Gentry, G. V. (1960). Differential effects of ionizing radiation upon "absolute" and "relational" learning in the rhesus monkey. *J. genet. Psychol.* **97**, 245.

Riopelle, A. J. (1962). Some behavioral effects of ionizing radiation on primates. *In* "Response of the Nervous System to Ionizing Radiation" (T. J. Haley & R. S. Snider, eds.), pp. 719–728. Academic Press, New York.

Riopelle, A. J., Grodsky, M. A., & Ades, H. W. (1956). Learned performance of monkeys after single and repeated X irradiations. *J. comp. physiol. Psychol.* **49**, 521.

Rogers, C. M., Kaplan, S. J., Gentry, G., & Auxier, J. A. (1954). Some effects of cumulative doses of X-radiation upon learning and retention in the rhesus monkey. USAF Sch. Aviat. Med., Proj. No. 21–3501–3003 (Rep. No. 11).

Ruch, T. C., Isaac, W., & Leary, R. W. (1962). Behavioral and correlated hematologic effects of sublethal whole-body irradiation. *In* "Response of the Nervous System to Ionizing Radiation" (T. J. Haley & R. S. Snider, eds.), pp. 691–703. Academic Press, New York.

Settlage, P. H., & Bogumill, G. P. (1955). Use of radioactive cobalt for the production of brain lesions in animals. *J. comp. physiol. Psychol.* **48**, 208.

Tinklepaugh, O. L. (1928). An experimental study of representative factors in monkeys. *J. comp. Psychol.* **8**, 197.

Warren, J. M., Kaplan, S. J., & Greenwood, D. D. (1955). The solution of discrimination-reversal problems by normal and irradiated monkeys. USAF Sch. Aviat. Med., Proj. No. 21–3501–3003 (Rep. No. 16).

Chapter 15

Field Studies

Phyllis Jay

Department of Anthropology, University of California, Davis, California

I. INTRODUCTION

Behavior of free-ranging primates has been the subject of many recent field studies, and as a result the comparative study of primate behavior is beginning to take form. Some generalizations are possible, and the trends apparent in recent studies can be evaluated.

In the nineteenth century, many notions about primate behavior were formed from anecdotes, incidental observations, and travelers' tales. Yerkes and Yerkes (1929) summarized this early era in their book, *The Great Apes.* Yerkes greatly influenced subsequent research on apes and monkeys by stimulating the field studies of Nissen, Bingham, and Carpenter. Zuckerman's *The Social Life of Monkeys and Apes* (1932) also has had a profound and continuing effect. As a result of the pioneering efforts of these men, the study of primate behavior began to be systematized, and today inaccurate data are rapidly being replaced by valid observations not only of monkeys and apes in the laboratory but also of free-ranging animals in natural field conditions. Many basic concepts such as range, dominance, and the existence of breeding seasons can now be reexamined. But the living primates are a large and widely diversified group of mammals comprising some 50 genera (see endpapers) and 600 species and named subspecies. Only a small percentage of primate species has been studied, so present generalizations will probably need to be modified as the behaviors of more species are observed.

In 1963, more than 50 workers from many countries studied primates in the field. These and numerous other recent studies of nonhuman primates have been tabulated by DeVore and Lee (1963). Recent studies range in length from brief single visits to long-term repeated investigations. Although preliminary field studies of many species are still needed, the trend is toward longer field investigations. Important seasonal, regional, and group variations have been recognized, emphasizing the need for long-term studies of several populations in different localities. The thorough understanding of the social behavior of a species requires a major cooperative effort extending over many years. The Japan Monkey Centre has observed semifree-ranging Japanese macaques (*Macaca fuscata*) for over 12 years, the longest study of any monkey or ape to date.

A. The Field as a Natural Laboratory

Learning, perception, affection, curiosity, etc., are important aspects of primate behavior, as many chapters in these volumes clearly demonstrate. But the adaptive functions of each of these aspects of behavior are clear only in a free-ranging social situation where the advantages of being an intelligent, perceptive, affectionate, curious animal can be seen as they cannot be in a laboratory or cage.

Most behavior in a free-ranging primate group is routine and stereotyped; life is simple and repetitive for monkeys and apes. Aspects of this repetitive behavior may be categorized easily by the observer, and such categories as play, sex, or dominance activity are common to all descriptive reports. The frequency of these behaviors not only varies greatly from species to species, but the amount of time devoted to each is altered by such unusual events as crowding, famine, artificial feeding, or captivity.

The basic daily routine of activity for monkeys and most apes is well known to group members. Morning activity periods are followed by midday resting and feeding (Fig. 1). A late-afternoon activity period, during which the group may move from one part of its familiar range to another, is followed by movement to the sleeping trees (Fig. 2). Some species, such as baboons (*Papio*), often sleep in the same area each night; hence the group may make a regular daily circuit away from and back to the sleeping trees (DeVore, 1962). The social organiza-

FIG. 1. Group of Ceylon toque macaques sits in the afternoon sun. Four adult females with infants of less than 1 month of age sit with three adult males (two on right and one on left) and one subadult male (facing observer). Notice the full cheek pouches of many of the animals. (Photograph by Phyllis Jay.)

Fig. 2. Group of northern Indian langur monkeys approaches trees in which it will sleep. Group moves leisurely, without any definite arrangement of group members. (Photograph by Phyllis Jay.)

tion of the group is familiar to each member, and well-established patterns of dominance provide a high degree of predictability of behavior for the animals. During the average day, the individual moves, grooms, eats, and rests with familiar animals in a familiar place. Novelty is minimal.

The sequence of normal events can, however, be interrupted. Crises do occur, and in order to survive the species must be able to cope with them. The normal, relaxed, stable, healthy group with a regular daily routine may experience changes in any of these qualities. Severe dominance fighting within a group because of too many adult males or an unbalanced sex ratio results in increased tension, wounds, and social stress. Males may fight over a limited number of receptive females and cause severe social disruption. Washburn and DeVore (1961) observed more aggression in baboon groups with a disproportionately high percentage of adult males.

Parasites and diseases are fairly common in wild monkeys and apes. Primates can survive even severe and numerous fractures that have repaired with differing degrees of success. Of 260 skeletons of gibbons (*Hylobates*) examined by Schultz (1956), 33% had healed fractures. High percentages of healed fractures are recorded for other apes and for several species of New and Old World monkeys. Schultz (1956) has summarized the data on other pathological and teratological conditions.

The full range of normal relationships of structure, function, and behavior can only be observed in a free-ranging group of primates. For example, it has been established in laboratory studies that the clinging reflex takes precedence over the righting reflex (see Chapter 8 by Harlow and Harlow). The great significance of this order of response is not clear in the laboratory but is immediately apparent in the field. The clinging response is an essential adaptation for maintaining contact with a moving and often very active mother. If the infant monkey is to survive, it must be able to cling to its mother when she moves. Also, in a situation where it is essential for the individual to keep up with the group, it would be difficult if not impossible for a mother to survive if she had to take full responsibility for supporting a helpless infant (Washburn & Hamburg, 1965a).

Similarly, the adaptive significance of many anatomical structures is not obvious in a cage. Old World monkeys sleep at night sitting on small branches away from the main tree trunk where they are safe from climbing carnivores that hunt after dark. Specialized sitting pads, ischial callosities, have evolved which allow the monkeys to sleep comfortably in an upright position even on very small branches (Washburn, 1957). The great apes, on the other hand, have evolved sleeping patterns in which they lie down rather than sit, and the percentage of animals with callosities is lower. These large apes build nests that allow them to sleep in trees, nests that function for these primates in the same way that callosities do for monkeys and gibbons, which do not construct nests. Field studies (Goodall, 1962, 1965) and laboratory research (Bernstein, 1962) indicate that nest-building is probably in part a learned ability. Washburn (1957) suggests that this is an instance of a learned pattern of behavior that has assumed the function of an anatomical structure, which in turn has gradually been lost.

B. Field Procedures

The field studies on which this chapter is based have been made in a wide variety of habitats and under very different conditions. In general, those species that live in thick vegetation are much more difficult to observe than are species in towns, villages, fields, and open plains. However, if study techniques are adapted to the situation it is possible to obtain data.

An understanding of the behavior of any animal ideally proceeds from the general to the very specific. Study of a primate should progress through three stages (Schaller, 1965a): (1) an ecological survey of as much of the primate's range as possible with emphasis on diversity of habitat, distribution and abundance of the species, and similarities or differences in group size and composition, food habits, etc., between populations of the species; (2) detailed observations of social life of a selected group, groups, or population, concentrating on the species' repertoire of behavior; (3) an intensive study of specific aspects of behavior, with field and laboratory experiments to clarify the complex relations of components of the behavior.

A reliable census of a species over a large area may be very difficult, since many primates are hard to locate in forests and dense vegetation. Some of the techniques employed in assessing populations of primates are recognition of individual groups (Carpenter, 1934, 1965), familiarity with many individual animals (DeVore & Hall, 1965; Jay, 1963a), observation of spoor (Schaller, 1963) and nest sites (Schaller, 1961), a survey of monkeys along transportation routes (Southwick *et al.*, 1961a, 1961b), and techniques of sample-area census that have been used in ungulate studies (Dasmann & Mossman, 1962).

Detailed description of the vegetation in which primates live is essential. Several species, such as baboons and rhesus, occupy a wide variety of distinct vegetation types, and it is essential to determine variations of behavior in different settings. However, a species can sometimes be observed in one type of vegetation, whereas it cannot be observed in other types that it also inhabits. This is true of the Indian langur (*Presbytis entellus*) (Jay, 1965).

Ideally, the presence of the observer should have no effect on the behavior of the group being observed, since the major objective of a field study is continuous observation of undisturbed animals. Often this means observing from considerable distances or from blinds such as Hall (1963a) used for baboons and Kortlandt (1962) for chimpanzees (*Pan*). Carpenter (1934) observed howler monkeys (*Alouatta palliata*) from blinds and by habituating one group to his presence. After several months of quiet, slow approach, Jay (1963a, 1963b, 1965) habituated one group of northern Indian langurs to her presence within the group. DeVore (1962) used a Land Rover to follow groups of savanna-living baboons. Other methods of habituation are discussed by Schaller (1965a) and Schneirla (1950). The Japan Monkey Centre feeds 15 groups of Japanese macaques at certain locations (Fig. 3). Feeding in restricted areas tends, however, to increase aggression while reducing grooming and play activity.

Since some behavior patterns observed in the field lend themselves to experimentation without excessive interference with the natural life of the animal, hypotheses about many social relations can be tested. For example, the Japan Monkey Centre observed the acquisition of food habits (Section II, B) by presenting new foods to groups and recording their acceptance by group members with different social status. Experiments may be used to test group and individual reactions to situations which normally occur very seldom. DeVore (1962) tested one aspect of dominance among adult male baboons by throwing food between two animals. Hall (1962) presented a snake and scorpions to baboons to see responses to these potentially dangerous animals.

A few observers have recorded the reactions of groups of primates to naturally occurring disasters, such as the flooding produced by the construction of the Kariba Dam which isolated baboons on an island with little food (Section III, A, 2, *a*).

II. ECOLOGY

A. Habitat and Distribution

Table I summarizes the habitats of the apes and of those monkeys for which information is available. Tappen (1960) has discussed in detail the distribution of African monkeys and Napier (1962) has discussed the habitats of monkeys. Both authors emphasize that a survey of primate genera must take into account use of vertical dimensions of the habitat as well as horizontal distribution.

A number of ecological features limit the distribution of arboreal mon-

TABLE I
Habitats of Monkeys and Apes

Species	New World	Africa	Asia	Primarily arboreal	Arboreal and terrestrial	Primarily terrestrial	Habitat
Spider (*Ateles geoffroyi*)[a]	X	—	—	X	—	—	High level of forest
Howler (*Alouatta palliata*)[a]	X	—	—	X	—	—	High level of forest
Capuchin (*Cebus capucinus*)[b]	X	—	—	—	X	—	Primarily arboreal
Mangabeys (*Cercocebus* spp.)[a]	—	X	—	X	—	—	High level of forest
Guenons (*Cercopithecus* spp.)	—	X	—	—	X	—	Varied habitats, more arboreal
Red-tail (*Cercopithecus nictitans*)[c]	—	X	—	—	X	—	Primarily arboreal
L'hoesti (*Cercopithecus l'hoesti*)	—	X	—	—	X	—	
Vervet (*Cercopithecus aethiops*)	—	X	—	—	—	X	Mostly terrestrial, widest distribution of all guenons
Patas (*Erythrocebus patas*)	—	X	—	—	—	X	
Chacma baboon (*Papio comatus*)[a, e]	—	X	—	—	—	X	Savanna
Olive baboon (*Papio doguera*)[a, e]	—	X	—	—	—	X	Savanna
Hamadryas baboon (*Papio hamadryas*)[f]	—	X	X	—	—	X	Desert
Drill (*Papio leucophaeus*)	—	X	—	—	—	X	Forest
Mandrill (*Papio sphinx*)	—	X	—	—	—	X	Forest
Gelada (*Theropithecus gelada*)[g]	—	X	—	—	—	X	Mountains of Ethiopia

Species					Environment
Macaques (*Macaca* spp.)[a]	—	X	X	—	Widely varying environments
Rhesus (*Macaca mulatta*)[h]	—	X	—	X	Widely varying environments
Guerezas (*Colobus* spp.)[i]	X	—	—	—	
Olive colobus (*Colobus verus*)[i]	X	—	—	—	Low canopy level of forest
Nilgiri langur (*Presbytis johni*)	—	X	—	—	Restricted distribution
Common langur (*Presbytis entellus*)[j]	—	X	X	—	Widest distribution of all langurs, most time on ground
Ceylon gray langur (*Presbytis entellus thersites*)[k]	—	—	—	—	Dry forests, wet forests
Purple-faced langur (*Presbytis senex*)	—	X	—	—	Mountain forests
Gibbons (*Hylobates* spp.)[l]	—	X	—	—	Rain forests
Siamang (*Symphalangus syndactylus*)[l]	—	X	—	—	Rain forests
Orangutan (*Pongo pygmaeus*)[m]	—	X	—	—	Rain forest
Chimpanzee (*Pan troglodytes*)[n]	X	—	X	—	Forest
Gorilla (*Gorilla gorilla*)[o]	X	—	—	X	Rain forest

[a] Napier (1962).
[b] Carpenter (1958).
[c] Haddow (1952).
[d] Hall (1963a).
[e] Chacma and olive baboons may possibly be two races of the same species (DeVore & Washburn, 1963).
[f] Kummer and Kurt (1963).
[g] Starck and Frick (1958).
[h] Southwick *et al.* (1961a, 1961b).
[i] Booth (1955, 1956, 1957, 1958).
[j] Jay (1963a).
[k] Phillips (1935).
[l] Carpenter (1940).
[m] Schaller (1961).
[n] Reynolds and Reynolds (1965).
[o] Schaller (1963).

Fig. 3. Large group of Japanese macaques assembles at a permanent provisioning place to eat sweet potatoes. (Photograph by Hiroki Mizuhara.)

keys, resulting in relative reproductive isolation of local populations differing in coat color and other features. First, the distribution of arboreal monkeys is restricted by open, relatively treeless areas. Second, rivers are barriers to arboreal monkeys but not to terrestrial forms, many of which swim. Third, the diet of most ground-living monkeys includes a wider variety of foods than does the diet of most leaf-eating arboreal monkeys (Section II, C). The major result of the greater diversity in habitats of ground-living genera is their wide and continuous geographical distribution, with less subspecific variation than among arboreal monkeys.

The notable exception to these tentative correlations between habitat, distribution, and speciation is found among the apes. The gibbon, the most arboreal ape, usually remains beneath the uppermost canopy of the forest. The genus is found throughout much of southeast Asia and is divided into a number of species. The orangutan (*Pongo pygmaeus*) is a single arboreal species living in mature forests of Borneo and Sumatra. The chimpanzee appears to be at ease both in the trees and on the ground. The gorilla (*Gorilla gorilla*), a single species, is extremely restricted in its distribution in part because of its diet, which is primarily leaves (Schaller, 1963). Although the gorilla climbs trees readily, it is the most ground-living of apes.

B. Home Range and Core Areas

Each group of primates lives in a distinct locality called its "home range." While this range can be expressed in terms of acres or square miles, it is more important to recognize home range in the context of the behavior of a species. Type of habitat, amount of space, pattern of use, and relations with other groups of the same and different species are integral facets of home range. To the extent that a primarily ground-living monkey sleeps in trees each night, eats in trees during fruiting and growing seasons, and depends on trees for safety, the home range of that species has three dimensions. Similarly the habitat of an arboreal monkey is also three-dimensional. The common northern Indian langur monkey, the most ground-living of the Asian langurs, is at home both in the trees and on the ground; never far from trees, it may nevertheless spend as much as 80% of the day on the ground (Jay, 1963a).

Although the home ranges of all monkeys have both horizontal and vertical axes, many arboreal species do not utilize all levels of the arboreal habitat (Napier, 1962). Some Asian langurs probably do not come to the ground and most species of African guerezas (*Colobus*) prefer levels above the shrub layer in primary forest. Howler monkeys also prefer the upper levels of primary forest and tend not to use secondary regrowth or scrub forest (Carpenter, 1934). Thus, some arboreal species may inhabit only segments or levels of the vertical axis and not venture above or below that level.

Among different species of monkeys and apes, home range varies in size from a few acres to well over 15 square miles. On the basis of available field studies it appears that terrestrial species have larger home ranges than do arboreal species (DeVore, 1963a). The arboreal gibbon, for example, lives in small groups whose home range totals approximately 0.1 square mile and often less. On the East African savanna of the Royal Nairobi Park each group of about 40 baboons typically occupies about 15 square miles during the course of a year (DeVore, 1963a). However, baboons living in forest areas probably have much smaller home ranges than do savanna groups. A group of approximately 17 gorillas normally occupies between 10 and 15 square miles in the region of the Virunga volcanos of Africa (Schaller, 1963). In the Budongo forest, range of a local population of chimpanzees is approximately 6 to 8 square miles, although groups of mothers usually travel much less than groups composed only of adult males (Reynolds & Reynolds, 1965). Goodall (1965) reported that many chimpanzees (*Pan troglodytes schweinfurthii*) in the Gombe Stream Chimpanzee Reserve in Tanganyika ranged over 80% of the 30-square-mile reserve and some individuals even left the reserve.

Population density varies greatly among primates and among areas supporting different types of monkeys and apes. While ground-living gorillas have population densities as low as 1 or 2 animals per square mile, it is quite conceivable that in forest areas inhabited by many arboreal species the total primate population may be as high as 200 individuals per square mile. On Barro Colorado Island, the density of howlers alone would account for one-third of this figure. In Africa, where one area of stratified forest can support three or four species (Napier, 1962), the total primate density is undoubtedly quite high.

Howler monkey groups on Barro Colorado Island counted in 1933 by Carpenter could not have had more than 0.2 square mile for exclusive use. By 1959 the amount of land available to each group was no more than about 0.14 square mile. The average population density among gibbons observed by Carpenter (1940) was 10 to 12 animals (approximately 2 to 3 groups) per square mile. Chimpanzees observed by Goodall had a population density of approximately 2.6 per square mile or, relative to land actually used, 3.3 animals per square mile. In the Budongo forest, the density was approximately 10 chimpanzees per square mile (Reynolds & Reynolds, 1965).

Areas of intensive use within a home range are referred to as "core areas" (Kaufmann, 1962). They are usually connected by traditional pathways. For langurs (Jay, 1963a) and baboons (Hall & DeVore,

1965), core areas typically contain sleeping trees, refuge sites, food sources, and water. One baboon group in Nairobi Park had a home range of over 15 square miles, but the three core areas contained in this large range accounted for only 3 square miles. Langur groups in forest regions of central India shift their core areas as the seasons change and different food plants become available.

In most species, contacts between groups usually occur in overlapping areas of home range. However, various spacing mechanisms, including vocalizations (Section III, B, 3), help keep groups apart, and aggression is rare. Home ranges of adjacent baboon groups may overlap about 50%. As one group's range may be overlapped by the ranges of several other groups, each group seldom possesses more than 1 square mile of exclusive range; hence, population density of baboons is much higher than a simple addition of home-range size would indicate (DeVore, 1963a). This is also true of forest langurs in northern India since as much as half of a group's home range may overlap other home ranges. Importantly, however, core areas within langur and baboon home ranges rarely overlap ranges of adjacent groups. Since many parts of the home range are seldom used and others are used only when the group is *en route* from one core area to another, there are few occasions when groups come in contact.

Adjacent home ranges of groups of one species are normally spaced in such a way that each group has access to adequate amounts of food and water. When drought or food scarcity makes it impossible for a group to survive within its range, sources of food and water may become overcrowded. Many baboon groups may concentrate around water holes during dry seasons with a minimum of aggression and then return to normal home ranges when the water supply is normal (Washburn & DeVore, 1961). In other instances when the crisis is prolonged and severe there may be a breakdown of normal social organization (Hall, 1963a). Similarly, rhesus macaques (*Macaca mulatta*) living in severely overcrowded conditions in Indian cities are aggressive, and fights between males of the same and different groups are frequent. Serious wounds often result, and old scars were visible on most adult animals (Southwick et al., 1965). This is the only evidence from recent field studies that any species of monkey or ape defends the boundaries of its home range against others of the same species. Burt (1943, 1949) and Bourlière (1956) have defined "territory" as the defended part of home range. According to this traditional definition, primates do not maintain territories under normal conditions. However, according to Pitelka (1959), the most important feature of territoriality is not that defense behavior is displayed to ensure control of an area by an individual or by a group, but that exclusive use of an area is maintained in some manner by its inhabitants. It is only in this sense that there is justification in applying the term "territory" to the normal use of space by the free-ranging primate groups that have been studied.

C. Diet

Diet is often equated simply with nutritional adequacy of food, but field observation suggests that diet is also important in species distribution and in ecological and behavioral interaction among species that live in the same area. All monkeys and apes eat fruit, buds, and tender leaves, but the proportion of the total diet formed by these foods varies greatly among different kinds of primates. Diet is more than a list of foods; it is an essential part of a complex of behavior, morphology, and physiology. Different diets make different demands on teeth and muscles of mastication (James, 1960), with consequent differences in the patterns of wear and use.

Diet can also set limits for exploratory behavior. The wider and more varied the diet of a species the more ecological niches and habitats it can investigate and live in. For example, the rhesus macaque's diet is much wider than that of the langur. Rhesus eat everything included in langur diet with the exception of mature leaves. In addition rhesus eat whole grains and a large number of roots and seeds that langurs never or seldom eat. Rhesus macaques often live in cultivated fields and in villages and cities where the adoption of new foods makes their survival possible. In contrast, Japanese research workers in India found it very difficult to induce langurs to take foods that did not normally occur in their diet, and in one area had to abandon their attempts (Kawamura, personal communication).

Observations of the Japanese macaque by members of the Japan Monkey Centre provide the most comprehensive account of food habits of any single species, as well as of differences in food preferences among different groups of one species (Kawamura, 1959). Each group of Japanese macaques eats foods not eaten by other groups. For example, the Minoo Ravine group removes earth from slopes to get the roots of a certain plant, whereas the Takasakiyama group does not do this even though the plant grows within the group's range. Some foods are eaten by several groups and not by others. Some groups raid rice-paddy, whereas others, also living near paddy, do not enter them.

New foods including sweet potatoes and candy have been deliberately introduced into a group's diet. "Acculturation usually starts among infants whose behavior is 'free-floating' and not well fixed" (Kawamura, 1959, p. 47). New food habits may originate with a single animal in the group (Imanishi, 1957). It is even possible to predict, depending on which monkey first acquires the habit, the social channels through which the new behavior pattern is passed throughout the group. Depending on the group, certain paths of transmission of new habits are more frequent than others. An infant often introduces a new habit into the repertoire of its mother, or else the patterns spread between especially intimate individuals.

Fig. 4. Adult female baboon pushes food out of cheek pouch with fist. (Photograph by Irven DeVore.)

The varied diet of baboons (Fig. 4) and their ability to dig below the surface of the ground for succulent roots have enabled them to move out onto the savannas. Most of the baboon diet consists of grass, buds, fruit, and plant shoots (Hall & DeVore, 1965), but insects, fledglings, birds' eggs, and meat are also included. DeVore and Washburn (1961) observed baboons killing and eating two newborn Thompson's gazelles (Fig. 5), two half-grown hares, and three nestlings of ground-nesting birds. On these occasions, the animals were killed and eaten when the baboons came upon them in the grass accidentally. There was no indication that baboons ever went out of their way to hunt or to capture

Fig. 5. Adult male baboon starts to eat a newborn Thompson's gazelle which it has killed. (Photograph by Irven DeVore.)

animals. Groups of baboons several times passed near carrion but made no attempt to eat any remains from carnivore kills.

Very little field data are available on ecological interactions of species that occupy the same area. Gibbons and siamangs (*Symphalangus syndactylus*) have been observed feeding together in the same tree in central Sumatra without any indication of aggression (Carpenter, 1958), and Schaller (1961) observed a group of gibbons less than 100 feet from two orangutans in Sarawak. The diet of the gibbon consists of fruits (80%) and leaves (20%), and may also contain insects, eggs, and nesting birds (Carpenter, 1940), giving it some latitude in habitat. Gorillas and chimpanzees live in the same forests in several areas of central and western Africa, but it has not been determined whether they are ecologically separated. Mountain gorillas in Congo basin rain forests and in the Kayonza forest of Uganda are mainly herbivorous, with fruit comprising a very small part of their diet (Schaller, 1963). In a similar forest habitat the chimpanzee diet is predominantly fruit. Reynolds and Reynolds (1965) estimated that 90% of the diet bulk of the chimpanzees living in the Budongo forest is fruit. Goodall (1963) also indicated that the bulk of the chimpanzee diet in the Gombe Stream Reserve is fruit. Thus, when gorillas and chimpanzees live in the same forest there may be little competition for foods. Northern Indian langurs and rhesus macaques occupy many areas together and here too, although there is an overlapping of diets, there is seldom competition for specific foods. I observed two adult rhesus macaques living in a langur group with no apparent difficulty (Fig. 6). Relations between langurs and other species living in the same area are usually peaceful (Fig. 7).

The chimpanzee, in addition to eating ants, termites (see below), and other insects, is the only nonhuman primate that regularly catches, kills, and eats meat. Goodall (1965) reported having observed chimpanzees in the Gombe Reserve eat meat on nine occasions. Once she saw a red colobus monkey (*Colobus badius*) hunted and killed by chimpanzees. One adolescent male chimpanzee climbed a tree near that in which the colobus was sitting, and then waited quietly. While the monkey watched this chimpanzee, another adolescent male chimpanzee climbed up to the colobus and caught it with its hands, presumably breaking its neck. The first adolescent male and five other chimpanzees, including a mature male and an older infant, then climbed to the animal holding the carcass. The colobus was torn into portions for each chimpanzee except the infant. On other occasions when Goodall observed meat-eating, the prey was first in the hands of an adult male. Other chimpanzees then sat close to him and held out their hands in what Goodall described as a "begging" gesture. Goodall also observed chimpanzees catch and kill a small bushpig and two young bushbuck.

The chimpanzee is the only nonhuman primate to use tools to get even a relatively minor part of its diet. For almost 9 weeks at the start of the rains, chimpanzees feed for 1 or 2 hours a day on a common species of termite, which open their nests to the surface and then seal the entrance with a thin layer of soil. Several weeks before the termite season actually starts, the chimpanzees examine the hills. As soon as a chimpanzee

FIG. 6A.

FIG. 6B.

FIG. 6, A and B. Adult rhesus monkeys that were full-time members of a group of northern Indian langurs. (A) This male rhesus (foreground) is dominant over the entire group, and when he passes adult langurs frequently stand and walk away. (B) This female rhesus (rear) frequently grooms and is groomed by langurs. (Photographs by Phyllis Jay.)

notices a fresh thin layer of dirt sealing up the hole, it scrapes it away and proceeds to get the termites. The chimpanzee does this by picking a grass stalk, a thin twig, or a piece of vine, and very carefully pushing it down the open hole using either hand. After a short wait the chimpanzee withdraws the stick, picking off the clinging termites with its lips. In picking off the insects, the chimpanzee may use the back of one wrist to support one end of the tool. If the stalk being used bends, the chimpanzee either uses the other end or breaks off the damaged portion. Should the selected stick be too long, the chimpanzee breaks off a piece. If it is too leafy, the leaves are stripped away with lips or fingers. Several times, when the nearest termite hill was at least 100 feet away and out of sight, a chimpanzee was observed to pick a grass stalk which it then carried to the termite hill and used as a tool. One male carried a grass stalk in his mouth half a mile, examining six termite hills, one after the

Fig. 7. Groups of southern Indian langurs and bonnet macaques often feed together peacefully (Photograph by Phyllis Jay.)

other, none being ready for working (Goodall, 1965). A species of ant which makes a nest in trees and one which makes an underground nest are caught and eaten by chimpanzees in a similar manner.

Beatty (1951) described chimpanzees in Liberia breaking open palm nuts by hammering them with rocks. Merfield and Miller (1956) described chimpanzees poking long twigs into the entrance holes of bee ground nests and then pulling out the twigs covered with honey.

Pitman (1931) observed free-ranging gorillas using a stick to get food which was out of reach. In over 12 months of observation Schaller (1963) did not observe any use of tools among the mountain gorillas.

Hall (1963b) commented on the apparent discrepancy between the potential for tool use that captive nonhuman primates show in situations designed to elicit it and their failure to exploit this potential in the wild. In captivity, species of great apes and monkeys learn to use sticks, sacks, boxes (Klüver, 1933; Bierens de Haan, 1931), and even live rats (Klüver, 1937) to haul in food objects that are out of reach. This skill is mastered spontaneously as well as through progressive training, and some species appear to have more aptitude than others in this regard.

Free-ranging individuals or groups are seldom called on to produce unusual forms of behavior, and yet the ability to cope with the unusual is an essential part of the patterns of behavior that are built into the species' repertoire. The potential variation in behavior of any species is not apparent from routine observation of usual daily events, but must be discovered in the laboratory.

D. Predators

Precautions against predators and rapid response to their approach are essential for survival, especially among ground-living species such as baboons which wander far from the safety of trees while foraging. Life on the ground increases exposure to predators in contrast to an arboreal way of life, in which the animals are either in large trees or within a few seconds of them. Once the monkey is in large trees, it can escape from ground-living predators, even climbing carnivores, by jumping from small branch to small branch. Arboreal monkeys need only descend quickly below the canopy layer of forest to be effectively out of the reach of raptorial birds.

All kinds of monkeys that have been observed give some alarm call which effectively warns any group members that have not observed a predator. If two or more species living in the same area are endangered by the same kind of predators, it is to their advantage to be able to respond to the danger signals of each other. For example, langurs respond to rhesus alarm calls (Jay, 1963a) as do chimpanzees to alarm calls of baboons, other species of monkey, bushbuck, and several species of bird (Goodall, 1965). A normally relaxed group becomes immediately alert when a predator threatens.

Reactions to predators differ among species. A baboon has little chance of escape if it runs from lions or leopards when the group is away from trees, and, therefore, when moving, the weaker, vulnerable females and young are surrounded by a formidable rank of large, powerful males. For baboons living on a savanna, survival can be equated with life in a group; the alternative is a brief solitary existence and probable death by predation (DeVore, 1963a, 1963b; Washburn & DeVore, 1961).

Japanese macaques display patterns of group organization similar to those of baboons (Imanishi, 1960). Although it has probably been generations since Japanese macaques have been subject to predatory action (Itani, 1954), forms of social organization which serve as effective defenses persist.

In response to an alarm bark, a group of Indian langur monkeys rushes into the nearest trees, where adult males bark at the predator while the rest of the group climbs high above them or moves back into the forest. In the initial flight each animal acts immediately and independently of other group members. Since langur defense is flight into nearby trees, there has been little selective advantage for the evolution of large males specialized to defend the group by fighting. As a result, among langurs as among other arboreal monkeys, there is much less sexual dimorphism than among the ground-living monkeys and apes (Jay, 1965).

Those anatomical features that equip the male monkey to fight and to defend weaker members of the group are developed to a striking degree among ground-living monkeys. They include large and robust bodies, large temporal muscles, and large canine teeth (Washburn & Avis, 1958). Similar morphological adaptations for life on the ground are also noted among the guenons (*Cercopithecus*), such as the vervet (*Cercopithecus aethiops*), that spend a lot of time on the ground, and the patas monkeys (*Erythrocebus patas*).

Among Old World monkeys for which data are available, there appears to be a positive correlation between adaptation for defense and the amount of agonistic behavior within the group. Usually, however, this aggressive behavior stops short of fighting, and fighting between groups is exceedingly rare. When it occurs, it is usually under conditions of abnormally high crowding such as among adjacent groups of city-living rhesus macaques (Southwick *et al.*, 1965).

Although in some species there is no clearly-defined male dominance hierarchy (Section III, C), this does not mean that males in these groups lack protective functions. Male howlers, vervets, and colobus have been observed to take action against potential predators; in other species males give alarm calls and threatening vocalizations whenever predators are nearby.

None of the apes appears to be subject to significant predation by carnivores. The arboreal gibbon is so agile in trees that it is improbable that any predator is a serious threat; Carpenter (1940) reported a complete lack of evidence of predation on the gibbon population he observed. Evidence of predation on orangutans is not available since no thorough study of these large tree-living apes has been undertaken. Neither Goodall (1965) nor Reynolds and Reynolds (1965) reported evidence of predation on the chimpanzee. Schaller discussed the evidence for predation on the mountain gorilla: "To my knowledge only one infant gorilla vanished from the groups which I had under frequent observation during the period of study, and the cause of its disappearance was not determined. In other words, I found no evidence that leopards habitually prey on gorillas. In an environment which contains a bountiful supply of defenseless prey, leopards need not venture combat with gorillas. It is possible that some leopards become gorilla killers just as some large cats show a preference for humans" (Schaller, 1963, p. 304). The sexual dimorphism and threat display (Section III, A, 3, *b*) of gorillas suggest a major adaptation to solve the problem of predators.

III. SOCIAL BEHAVIOR

A. The Group

1. THE GROUP AS A SOCIAL UNIT

Monkeys and apes live in social groups usually including members of both sexes and many ages (Section III, A, 2). The social boundaries of the group may be more or less permeable to strange or outside animals (Section III, A, 3), but within the group the members know each other well. Within this social environment of familiar animals, an important feature of life is the high degree of predictability of behavior of group members. Interactions are frequently determined by the participants' positions in the group dominance structure (Section III, C). Patterns of behavior have evolved which maintain a cohesive group and meet the primary adaptive tasks of the species, such as those of defense against predators (Section II, D), reproduction, which includes copulation and care of young (Section III, D), and obtaining an ample diet throughout the year in spite of seasonal variations in food supply. In some species, particularly terrestrial monkeys, group living has been the only efficient mechanism to meet these needs. A lone individual certainly could take care of some of these needs, but it would then lead a competitive and often dangerous life.

Group living assures fulfillment of many needs by what Washburn and Hamburg (1965b) have called multipurpose mechanisms. Attachments (responsiveness to other members of the species) and learning are effective means of assuring that the primary adaptive tasks will be accomplished. Since groups of monkeys often contain many members, an important "multiplier" function is an advantage of living in a large group. The individual monkey benefits from the alertness, protection, companionship, and attentions such as mutual relaxed grooming which can be derived from living with others. Since the members of monkey and probably most ape species are usually in sight of a familiar conspecific, the individual is also apt to learn and benefit from the innovations of other animals in the group.

Since a group of monkeys usually acts as a unit, the drives or desires of an individual member frequently may not be satisfied as soon as they arise. For example, a group of baboons crossing open savanna to get to water contains many thirsty monkeys. But each member of the group waits until the leading males have investigated cautiously whether it is safe to drink. The individual, although fully aware of the location of water sources, does not wander off to drink whenever thirsty, but instead waits for the entire group. In areas such as Victoria Falls where small pools of water are abundant, individuals can drink frequently and need not wait for the whole group to do so (Washburn & Hamburg, 1965a).

The traditional definition of a primate group as consisting of stable membership of both sexes and all ages must be modified to include the more fluid social relations of orangutans and chimpanzees (Section III, A, 3, b). Although individual chimpanzees move frequently from one aggregation of animals to another (Goodall, 1965; Reynolds & Reynolds, 1965), members of the local population probably know each other. With additional observation it will probably be possible to designate the local population of chimpanzees, spread over many square miles, as the basic unit of the species.

2. Group Size and Composition

a. Monkeys. Rhesus, Japanese, and bonnet macaques (*Macaca radiata*), baboons, langurs, and other monkeys may live in groups of 50 or more members of both sexes and many ages (see Fig. 1). In general, the more arboreal Old World monkeys such as Asian langurs and African colobus (Booth, 1957) tend to live in smaller groups than do ground-living monkeys such as baboons and macaques. In some species such as rhesus macaques (Southwick *et al.*, 1965), Japanese macaques (Imanishi, 1960; Mizuhara, personal communication), and howler monkeys (Carpenter, 1934; Collias & Southwick, 1952), peripheral animals on the edge of the group are observed.

Savanna baboon groups vary in size from 5 to 150 animals (DeVore, 1963a). Among most African baboons, local ecological conditions have important influences on group size. In the Drakensberg Mountains of South Africa, on the basis of a brief sampling of the baboon population, there appears to be a negative correlation between the altitude of a group's home range and the number of animals in the group. Hall (1963a) has suggested that this correlation depends on the availability of food. He has also estimated that average group size is lowest in arid Southwest Africa and highest in Southern Rhodesia.

When the Kariba Dam was constructed, a large island was formed in Lake Kariba, Southern Rhodesia. This 10-square-mile island offered very little food for baboons, and Hall (1963a) estimated that the island could adequately support about 15 to 30 animals. At the time of his observations, however, the island's population of refugee baboons was much larger than that and many had died. The surviving individuals were fully occupied in trying to sustain life on the overcrowded island. Hall observed one group of 31 and another of 17 animals, but most baboons wandered singly or in small groups of from 2 to 9. There was evidence that the social groups had gradually broken down into these smaller units under the stress of severe food shortage. These extreme conditions profoundly changed most of the normal behavior patterns of the island population. No females were in estrus, either fully or partially, nor was mounting or presenting observed. Only one infant was carried in the entire island population. Many individuals looked lethargic and no squealing or sounds of dissension could be heard on any part of the island. All activity was concentrated in seeking food. Fighting was not observed though there were several instances of lunging and chasing among adult males. The imbalance between ecological factors and population density had created severe disintegration in social organization.

The hamadryas baboons (*Papio hamadryas*) observed in Ethiopia by Kummer and Kurt (1963) have a group structure very different from that of other baboons. In 23 population samples, there were 18% adult males and 33% adult females. The size of sleeping groups was estimated to range from 12 to 750 animals, with a mean of 100 animals. The animals slept on rocks from 12 to 20 meters high with the number sleeping on one cliff changing every night. When the cliffs were crowded, it was almost impossible to tell which females belonged to which males. In the morning the large group left the cliff as a unit, but soon split into smaller units of from 5 to 40.

The basic daytime unit consisted of from one to four females and their offspring following a single adult male. Members of a one-male group rarely spread over more than 5 square meters. Young adult males constantly watched the females associated with them. If a female moved away, the male of her group threatened her. If she hesitated, he attacked her, bit the nape of her neck, and forced her to follow him closely. When food was concentrated in a small area, several groups might mingle until males began to bite the napes of their females and to threaten each other. Then the groups segregated, females crowding together in the "attack shadow" of their leader male. But contact among adult males was rare and no dominance order among males was observed.

Captive hamadryas baboons described by Zuckerman (1932) also formed one-male groups. Thus, such groups appear to be basic to the species, rather than the result of present ecological circumstances.

b. Apes. Although social groups of both gibbons and orangutans seldom exceed 5 individuals, gorillas and chimpanzees live in larger units, of from 12 to 20 for gorillas and from 30 to 60 for chimpanzees.

Carpenter (1940) surveyed 21 groups of gibbons in Siam. These groups totaled 93 individuals, including an equal number of adult males and females. Group size was typically between 2 and 6, usually consisting of 1 adult male, 1 adult female, and young. Of the 21 groups only 2 contained 2 adult males; in each of these, 1 of the 2 males was aged. Only 1 of the 21 groups had more than 1 adult female. Several males lived alone.

During his brief survey of orangutans in Sarawak, Schaller (1961) observed orangs traveling in small groups of 2, 3, or 4 animals. The range of group size which Schaller recorded was from 2 to 6, but many of the adult males in the population stayed alone. The most frequently observed groups were composed of 1 adult male and 1 adult female, or subadults, or adult females with 1 or more subadults.

The average gorilla group is composed of 17 individuals (Schaller, 1963), but the number reaches as high as 30 for short periods. The group is usually contained within an area with a diameter of about 200 feet. A gorilla group typically contains both sexes, and young as well as adults. Lone males leading a solitary life for a month or longer are common, whereas lone females were not observed.

The chimpanzee population that Goodall (1965) observed contained between 60 and 80 individuals. Animals that live at opposite ends of the range may not come in contact with one another, but there is a great deal of contact among most individuals in the area. Goodall recorded some instances of chimpanzees "visiting" from outside the "resident" population. Adult males were frequently apart from groups and sometimes remained alone for several days. Adult females, on the other hand, were observed alone only eight times during the entire 2-year period of observation. It is unlikely that any individual chimpanzee remains outside of some group for as long as a month (Reynolds & Reynolds, 1965).

In the Budongo Forest, local populations may consist of from 60 to 75 chimpanzees (Reynolds & Reynolds, 1965). Here too, composition of small groups constantly fluctuates and the members of a "community" are never all together in a single place. Small groups move from one food source to another and tend to combine at certain times, as when there is a concentration of fruit. Feeding groups may then consist of from 15 to 50 animals, whereas at other times they may dwindle to only 3 or 4. Reynolds and Reynolds described four general types of small groups: adults of both sexes (no mothers with dependent young), adult males, mothers with young and other adult females, and mixed groups of both sexes and all ages. The many small, adventurous groups of mixed adults or of only adult males constantly explore for new food sources and direct groups with young to food concentrations by loud noises and calling (Reynolds & Reynolds, 1965). Because groups including young tend to move more slowly, the actions of the more mobile bands tend to maximize use of available forage.

3. GROUP STABILITY: THE SOCIAL BOUNDARIES OF THE GROUP

a. Monkeys. The membership of social groups of most species of monkeys is relatively constant. It is an exceptional situation in which the traditional group of familiar animals of all ages and both sexes breaks down or in which the local population of a species lives in social units that are smaller or of a composition that is not typical for the rest of the species. Langurs (Jay, 1965), baboons (Hall & DeVore, 1965), macaques (Altmann, 1960, 1962; Imanishi, 1960; Mizuhara, personal communication; Southwick *et al.*, 1965) and howler monkeys (Carpenter, 1965) have been described as living in "closed" societies: social groups whose membership is stable over long periods of time.

The langur monkey lives in social groups of stable membership which are hostile to strange or nongroup males. However, a few adult males leave the group to live alone or in groups composed only of adult males (Jay, 1963a, 1965; Kawamura, personal communication; Ripley, personal communication). Male groups of as many as 10 adults are probably composed of males from several bisexual groups in an area. It is not known why these males leave the group, but they clearly are highly motivated to regain membership, and they may attempt to do so at the risk of severe injury. On several occasions I observed individual northern Indian langur males trying to associate with a bisexual group. These males were always fought off and were not allowed to remain near the group. The only serious fighting observed among langurs was during these attempts of males to rejoin groups.

Groups of baboons also do not readily admit strangers. In more than 1,400 hours of observation of more than 25 troops of baboons, DeVore and Washburn (1963) recorded only two individuals changing groups. Although Carpenter (1942) cited examples of individual estrous rhesus macaques that moved temporarily from one group to another in the colony on Cayo Santiago (near the eastern coast of Puerto Rico), a later study by Altmann (1962) indicated that after the animals had organized into stable groups almost no individuals changed. Imanishi (1960) described the Japanese macaques studied by the Japan Monkey Centre as having closed social groups.

Prolonged observation of the Japanese macaque by members of the Japan Monkey Centre has produced detailed descriptions of group division. By the spring of 1959, when the large group of Japanese macaques at Takasakiyama had grown to 500 individuals, a 70-member splinter group was establishing itself (Sugiyama, 1960). During the fissioning process several fights were recorded between young adult males of the two groups. Females, however, displayed very little antagonism. Three young adult males from the original group formed the nucleus of the new group and one of them emerged as the leader. None of the leaders or "subleaders" of the original group joined the insurgents. Later this group was joined by 20 to 30 individuals, including newborn infants. This split probably was generated by overpopulation of the feeding area and the tremendous increase in size of the initial group. Group division of a different nature has been described by Furuya (1960) for the group at Gagyusan. Kawai (1960) has described the process of group formation in recently released macaques.

b. Apes. The gibbon and the gorilla have relatively stable and cohesive groups which stay together for several months, but the orangutan and chimpanzee are characterized by unstable social groups whose membership is in constant flux.

The membership of gibbon groups is probably constant over long periods (Carpenter, 1940). Although there is some competition for food, neighboring groups intermingle for short periods of time. Unfortunately, nothing is yet known about the social behavior of the siamang, which is closely related to the gibbon.

A comparison of gorilla and chimpanzee social organization indicates major differences between these two great apes. Gorilla groups are relatively permanent, most members staying together for months (Schaller, 1963). In contrast, the composition of a chimpanzee band is so extremely fluid that it may change daily or even hourly (Goodall, 1965). In a local population of gorillas, there are changes in group composition though not as frequently as among chimpanzees. In both kinds of ape, adult males are the most mobile and they change groups more frequently than do members of any other age-sex category. While some male gorillas stay with a group of females and young for over a year, other males move in and out of groups. If a male comes into a group with sexually receptive females, he is not repelled or chased by the already established males. There is some variation among gorilla groups in the degree of tolerance of outsiders. Intergroup relations are usually peaceful (Schaller, 1963). However, when two gorilla groups come together, the adult males may charge abruptly and sometimes may beat their chests. After this dramatic and unique display the groups usually settle down and feed quietly (Schaller, 1963, 1965b). Small bands of chimpanzees mingle with excitement, leaping about and vocalizing noisily before they quiet down and feed together (Goodall, 1965; Reynolds & Reynolds, 1965).

B. Communication

The social structure of a primate group is maintained by a complex and richly varied system of both gestures and sounds. Human observers tend to exaggerate the importance of sounds and underestimate the importance of gestures since sounds carry most meaning in human communication. The following discussion of communication among monkeys and apes is based primarily on Marler (1965).

In order to systematize and analyze the meaning of communication patterns, sounds and gestures must be recorded in a wide range of carefully observed social situations. A communicatory signal is produced by one animal and evokes a response in another. Factors either external or internal to the signaler, or more likely a combination of both, influence the production of a signal. To the extent that the surroundings of the signaler are also perceived by the receiving animal, they can modify the response given to a signal. So the signal cannot meaningfully be separated from the specific circumstances surrounding signaler and recipient. Furthermore, the immediate surroundings must always be viewed within the framework of the natural environment of the species.

Studies of monkey and ape communication have largely focused on description of signaling behavior with minimal information concerning responses evoked by the signals. Judgments of what constitutes signaling behavior have been subjective and will probably need revision in the future. Signals have been described and classified "typologically" or "syntactically" (Cherry, 1957; Marler, 1961). On the basis of descriptions, predictions are made concerning the potential properties of signal systems in communication. The need for distance signals varies with the system of social organization characteristic of the species. Signals effective for close-range communication are emphasized among species and in situations where the social group is normally contained within a small area. Vision and audition are relied upon most often for distance communication, whereas the sense of touch is important for communication at close range. According to Marler (1965), composite signals are overwhelmingly important, for in most situations what passes from one animal to another is not a single signal but a complex of visual, auditory, tactile, and sometimes olfactory signs.

1. OLFACTORY SIGNALS

Among primates, only the prosimians rely on olfaction as an important means of communication, in contrast to the limited use of olfactory signals in communication among monkeys and apes. Olfaction is possibly an important factor in sexual behavior among some species of macaques and baboons (Section III, D, 1). Olfactory cues may perhaps help in recognizing individuals.

2. TACTILE SIGNALS

Tactile communication among monkeys and apes is clearly very important. Mutual grooming is the most widespread and characteristic behavior pattern involving tactile stimulation (Fig. 8). The hand is

FIG. 8A. Adult female northern Indian langur grooms one of the most dominant adult males in the group. Several other langurs sit nearby. (Photograph by Phyllis Jay.)

FIG. 8B. Adult female baboon grooms an adult male while another adult female sits nearby with her infant. (Photograph by Irven DeVore.)

used to initiate physical contact in all species observed. Teeth, tongue, and nose are also used in many contexts, as well as the general body surface and the pelvic area. Tactile signals have been characterized and classified by the effectors used in signaling and the locus of stimulation of the recipient. Marler (1965) describes these as the "qualitative" properties of tactile stimuli and notes that the "quantitative" properties include intensity, completeness of the signal, and temporal pattern of delivery. For example, aggressive biting between two adult males in a fight is violent, rapid, and complete, whereas a male biting a juvenile in play is nonviolent, slow, and incomplete. Statistical data on the sequential patterning of behavior that precedes and follows tactile signals will aid in understanding their information content. Altmann (1962) has collected such data for rhesus macaques on Cayo Santiago.

3. AUDITORY SIGNALS

Auditory communication includes both vocalizations, generated by specialized vocal organs, and nonvocal sounds generated in many different ways. The inventory of nonvocal sounds with communicatory properties may include noises produced by locomotion and the striking of objects. Respiration can generate sound, and the sudden expulsion of breath can alert nearby animals to the presence of an individual. Many species of monkey and ape shake branches in aggressive displays (red spider monkeys [*Ateles geoffroyi*], Carpenter, 1935; baboons, Hall, 1962; Hall & DeVore, 1965; chimpanzees, Reynolds & Reynolds, 1965; gorillas, Schaller, 1963). Chimpanzees may drum on ground or trees, and gorillas make a similar sound by beating the chest with partially cupped hands (Goodall, 1965; Reynolds & Reynolds, 1965; Schaller, 1963) Laryngeal air sacs that amplify loud sounds are found in some primates (Andrew, 1963). A system of vocal signals may include either discrete or graded sounds. One probable source of error in counting vocal signals has been the subdivision of one continuously variable mode into discrete categories. Terms used to label signals should be, but have not always been, descriptive of the signal or its physical structure rather than its function. The size of the repertoire of vocal signals varies among different monkeys and apes. On the basis of available data, however, it appears that repertoires of approximately 10 to 15 basic sound-signal types are characteristic of the nonhuman primates (Marler, 1965). Langurs have two graded systems, the grunt-bark and the squeal-scream (Jay, 1965). The vocal behavior of rhesus macaques has been studied in detail (Altmann, 1962; Andrew, 1962; Rowell, 1962; Rowell & Hinde, 1962). The rhesus repertoire includes a wide range of sounds intermediate to the main categories (Rowell & Hinde, 1962). Rowell (1962) showed that nine sounds are constituents of one system linked to a continuous series of intermediate sounds. She also demonstrated correlations of these sounds with continuously varying social and environmental situations.

Many species of the arboreal guenons of Africa share their relatively densely populated habitat with other monkeys (Section II, B, 2), a situation which provides many possibilities for interspecies confusion of vocal signals. However, the limited data for some species of guenon suggest that communicative sounds among these species are highly structured and differentiated. This structuring may serve to maintain species specificity of sound signals for most of the vocal repertoire. Marler (1965) suggests that there is a correlation between degree of definition of structure of sound signals and degree of isolation from other species of similar size and structure. Relatively isolated species such as rhesus monkeys, baboons, chimpanzees, and gorillas have poorly defined sound signals.

Some vocal signals appear to be universal to monkeys and apes. Barks are used to signal alarm by gorillas, chimpanzees, baboons, rhesus, and langurs. Screeching and screaming sounds are signs of distress among all species observed. Other sounds are distributed widely but not universally through many species. Interspecies communication plays an important role, especially in giving and receiving danger signals (see Section II, D).

Vocalizations are an aid to spacing groups in some species of both New and Old World monkeys as well as several kinds of apes. Carpenter has suggested that among both gibbons and howler monkeys loud calls during vocal "battles" take the place of physical aggression. Carpenter (1940, p. 156) described two gibbon groups vocalizing louder and louder as they drew close together and finally mixing with no fighting after they ceased to call.

Adult males of the common northern Indian langur give a deep, resonant "whoop" call in the morning just before and while the group moves from sleeping trees, during the day when the group moves suddenly or for a long distance, and in early evening when the group moves toward the trees it will use for the night. These whoops carry long distances and probably indicate group position. Whooping occurred when troops came within sight of one another. When groups came in contact, which seldom happened, there was no fighting or aggressive behavior (Jay, 1965).

Ullrich (1961) described similar types of calls given by the African black-and-white colobus monkeys (*Colobus polykomos*). As among langurs, colobus calls do not instigate group interactions but are likely to serve as location calls among groups. It seems apparent that calls heard over long distances reduce the chances that adjacent groups will meet unexpectedly in the forest.

Chimpanzees are exceptionally noisy. One type of call, prolonged and high-pitched hoots, may be heard for a distance of as far as 2 miles. In the Budongo Forest of Uganda, chimpanzees are not very vocal during periods of food scarcity, but when food is abundant they make much more noise. Groups have been heard to hoot in bouts several minutes long repeated at intervals of less than an hour throughout the day. When a group starts to call it is frequently answered by other nearby groups. Reynolds and Reynolds (1965) suggest that this is one effective way of alerting other chimpanzees to areas of abundant food. During the fruiting season in one study area, an increase in drumming and noise-making pulled together what were otherwise two communities. Goodall (1965) has commented on the sharp increase in noise and activity when more than 10 chimpanzees are together.

The gorilla, on the other hand, is silent and vocally undemonstrative. Schaller reports no counterpart among gorillas to the noisy vocal display of an entire local chimpanzee group.

4. Visual Signals

Since most primates are mainly diurnal, they readily use and depend upon visual signals. "The close-knit social groupings of the higher primates, with the individual members in direct view of each other much of the time sets the stage for the development of the extraordinarily subtle and elaborate systems of visual communication which are revealed in every new study which is made" (Marler, 1965). A complete description of communication would include posture, movement, and physical characters that permit identification of the signaler. This is complicated further because nearby individuals may influence the original signal and the response of the recipient. Such influences have been observed in rhesus macaques (Altmann, 1962) and in baboons (Hall & DeVore, 1965).

If the animal is visible when signaling, any part of its body may be communicatively significant. Thus, the entire silhouette is significant among some East African guenons (Haddow, 1952), and a silver back is important among gorillas (Schaller, 1963). The highly colored, swollen sex skin of estrous females of most macaques, mangabeys (*Cercocebus*), baboons, and chimpanzees (Zuckerman, 1932) is a visual signal to males.

Facial expressions are probably the most important signals given by monkeys and apes (van Hooff, 1962). Both eyes and mouth are essential components in most gestures (Fig. 9). Direct gaze is often sufficient to assert dominance over a subordinate among gorillas (Schaller, 1963,

1965), baboons (Hall & DeVore, 1965), rhesus macaques (Altmann, 1962), bonnet macaques (Simonds, 1965), and langurs (Jay, 1965). Eyes are surrounded by conspicuously colored skin in many species, including mangabeys, guenons, and guerezas (photographs in San-

FIG. 9. Adult male Japanese macaque threatens the observer. (Photograph by Hiroki Mizuhara.)

derson, 1957), but van Hooff (1962) notes that there is usually relatively little movement of these parts. In contrast, the mouth may assume many significant shapes; variables include the degree of jaw opening, lip retraction, etc. (Hinde & Rowell, 1962). Ears can be moved back against the head or held erect. A wide variety of facial expressions and elaborate adornment of the face in many species combine to produce many subtle variations in visual signals, which probably have communicatory significance to the receiver. A wide variety of dynamic postural displays including the mode of carriage may change with the social situation and be effective visual signals. An example of this is the confident, relaxed walk of dominant males (Fig. 10) reported in most studies (see Section III, C).

FIG. 10. Dominant adult male Ceylon gray langur leads his group across an open field, grimacing and displaying his canines. Tail carriage differs from that of Indian langurs. (Photograph by Phyllis Jay.)

Visual signals may be and usually are constellations of most of the elements listed above. Individual components of the total signal can vary independently to produce a large range of possible combinations. Some elements, however, are not independent and occur in regular groups. Marler (1965) has suggested a tentative representative list of clusters of visual signal elements characterized by functional aspects of the situation in which they are used.

C. Dominance

Among all the primarily ground-living monkeys that have been observed, male dominance is probably the most important influence in social cohesion and regulation. In tree-living species, dominance status is also extremely important in regulating group activity, although dominance relations are usually less obvious and less frequently asserted than among terrestrial monkeys. Even among young baboons and those species of macaque for which field data are available, dominance patterns are more pronounced than among young langurs. It is usually impossible to determine precise patterns of dominance for young langurs, whereas among older juvenile baboons such patterns emerge clearly.

Dominance is expressed in a wide variety of forms from aggressive fighting to extremely subtle movements, as well as in leadership patterns and many other daily social activities. Dominant primates are able to secure desirable food, right of way on paths, grooming attention, and consortship with estrous females. Individuals may be more dominant in one type of situation than in others, although this may reflect motivation at that particular time more than ability to control the situation. These facets of dominance have traditionally been lumped into one broad category, obscuring major differences in types of interaction as well as important differences among species.

The great variety in dominance expressions and in the clarity with which the dominance structure emerges make it extremely difficult to compare species, so statements that one species is more or less dominant than another are often misleading. Differences in frequency of aggressive interactions have been reported among groups of the same species (Hall & DeVore, 1965) and among different species (Jay, 1965). If number of wounds and scars is a reliable measure of aggressiveness, it is clear that rhesus macaques, for example, are far more aggressive than langurs, and that city-living rhesus are more aggressive than forest groups. Dominance interactions are certainly more frequent among rhesus regardless of location than among langurs. Also, the percentage of subtle gestures in langur dominance interactions appears to be higher than among rhesus.

As a result of stable dominance structures in natural groups, the responses of each individual in a social situation are highly predictable. Because the members of a free-ranging group know each other well, "problems" such as determining initial dominance status of new animals seldom arise. In contrast, dominance patterns investigated in laboratories are often those displayed by animals put together as relative strangers in combinations usually not exceeding two or three individuals. It is evident from field studies that the presence of other individuals may influence status, and that the dominance system of the group is a combination of individual status and status in association with others. Dominance has not yet been observed in a laboratory in its total social context among many animals of both sexes and all ages.

Extent and kind of participation in activities such as food-getting, defense, reproduction, and leadership is often related closely to the individual's dominance status. However, dominance has frequently been measured in the laboratory merely by seeing which animal takes food. Field research indicates that it is exceedingly important to observe how aggressive interactions are *avoided* as well as how they are resolved. Avoidance of certain animals in the group obviates the necessity for interaction, a situation often not reproducible in laboratories where space is limited and there are no rocks or leafy boughs behind which an escaping animal can move to be out of sight.

Laboratory analyses of the components of dominance are essential, however. Influence of hormonal levels, wakefulness, and physical condition can be assessed accurately only under carefully varied and controlled conditions. Some experimentation is practical and desirable in the field, but the fine details of physiology need to be assessed in a carefully planned laboratory colony where the essential elements of social life and environment can be reproduced.

Dominance status is related to other variables besides size and strength even for a male, although these factors are important. Behavioral traits and responsiveness to certain situations and stimuli are also significant. Adult langurs differ strikingly in manner and rapidity of response to annoyance and threat, and in general the most dominant males are usually more relaxed and slower to respond to minor annoyances than are males of medium or lower dominance status.

The following is a brief summary of dominance patterns in several species of monkeys and apes.

1. Baboon

The baboon group is organized around the dominance hierarchy of adult males (Hall & DeVore, 1965). Among savanna baboons, dominance is not organized around one-male segments of the group as it is among hamadryas baboons, nor is it related to feeding since the group is normally very spread out and few animals eat near one another. Relative status among males is frequently expressed by dominance mounting during which dominant animals mount subordinates much as males mount receptive females. However, there is no intromission and seldom are there any pelvic thrusts. Subordinate animals often present their hindquarters to dominant animals although the latter may not respond by mounting. Sometimes the dominant animal only starts to mount and then walks away.

A novel situation confronting a savanna-living baboon group may be investigated first by the subordinate adult males, which are in the vanguard of the moving group, before the rest of the group approaches (DeVore, 1962). This behavior is adaptive, since these are the most expendable adult males. Their death probably would not affect the group birth rate since dominant males normally copulate with estrous females at the time ovulation is most likely to occur (Section III, D, 1).

The nature of the dominance hierarchy varies according to the size and composition of the baboon group. The simplest form of dominance organization is observed in smaller groups of from 15 to 35 members in which there is usually one completely dominant male. In one group of chacma baboons (*Papio comatus*) that lived in the Cape in South Africa there were three adult males, one of which was clearly the leader. This baboon was involved in more dominance interactions than any other in the group. In disturbing situations involving other animals, this male threatened the offender, and mothers with very young infants clustered near him for protection. In keeping with his superior dominance status, subordinate gestures were often directed to him. Because he had the prerogative of exclusive mating with most of the estrous females, he was probably the progenitor of most of the offspring in the group.

In another group of baboons in Kenya, six of the adult males were about the same size. Dominance interactions were more frequent in this group than in other nearby groups, and the hierarchy was not linear. Some of these six males were constant associates and consistently supported each other in dominance interactions. Several were so close that they never acted independently. Three of the six males formed what DeVore (1962) termed a "central hierarchy" which was completely successful against the three other adult males, which rarely formed coalitions. A male's dominance status is usually a combination of his individual rank (based on fighting ability) in a linear assessment of the group, and of his ability to enlist the support of the other males in the central hierarchy. A linear hierarchy usually was observed only as a temporary stage in group organization. DeVore also noted that the relative dominance status of groups depended on the dominance of the prominent males within them.

Transferred threat (that is, a threat directed toward a subordinate animal in response to a threat from a more dominant animal), as described by Altmann (1960) for the rhesus macaque, is also important in baboon social life. This form of redirected aggression can initiate chain reactions of dominance which usually start from the higher dominance levels and travel down the hierarchy to subordinate monkeys and finally to juveniles. Hall and DeVore (1965) suggested that this is an effective means of reinforcing dominance patterns. Transferred threats are less frequent in langurs among which dominance structure in general is less clearly defined and rigid than that of savanna baboons. These differences reflect the fact that langurs depend on nearby trees for safety, whereas baboons, on their daily circuit out on the savanna, must rely on elaborate social defenses.

Fig. 11. Adult male baboon displaying canines in a yawn threat. (Photograph by Irven DeVore.)

Adult male baboons and langurs harass consort pairs. Among baboons this involves pacing near the pair, teeth grinding, yawning (a canine display; Fig. 11), and finally chasing, attacking, or fighting. Harassment typically occurs when the adult male : female ratio in a baboon group is greater than the normal 1 : 2 or 1 : 3 and competition for females thus is greater (Washburn & DeVore, 1961). Harassment never occurred over food. Among langurs harassment is directed only toward consort pairs and the sequence rarely ends in fighting. Less dominant males run about a copulating pair barking, threatening, and slapping at the consort male. Aggressive behavior by the harassing male is directed almost exclusively to the consort male and not to the estrous female (Jay, 1963d).

2. Macaques

The highly aggressive behavior of free-ranging rhesus groups in a northern Indian city has already been mentioned (Section II, B). The bonnet macaque, observed by Simonds (1965), appears to be some-what more similar in this respect to the closely-related rhesus than to tree-living monkeys such as langurs. One subadult male was killed and several wounded in a dominance fight involving nearly all adult males in the group. Of 11 males in one bonnet macaque group, 5 were definitely dominant and policed the group, determining most daily activities. Four less dominant males tended to avoid these dominant males. In general, however, bonnet macaques appear to be less dominance-oriented and aggressive than their northern relatives, the rhesus. In sharp contrast to rhesus, bonnet males groom each other frequently. Simonds also noted that "adult male bonnet macaques play regularly" with infants, juveniles, and other fully adult males—a pattern not observed in any other monkey species.

The Japanese macaque social group is organized in dominance hierarchies reflected in the spacing of group members as well as in daily activities. The center of the group is occupied by the most dominant male(s) and dominant females with infants. In concentric circles around this central core are the other members of the group, with the least dominant of both sexes in the outermost rings. Kawai (1958) distinguished two types of rank among Japanese macaques. Dependent rank is influenced by the presence of other animals; basic or independent rank is not. Dependent rank appears first in an individual's life, as the rank of an infant Japanese macaque is greatly influenced by the rank of its mother. In general, it appears that the social status of the mother is of far less importance to the young in arboreal species than in terrestrial species where the associations and development of the infant may be substantially affected by the mother's status.

The Minoo-B group of 29 Japanese macaques described by Kawamura (1958) was led and dominated by a female. This, however, was the only instance of female leadership reported when males were present in the group.

3. Langurs

The northern Indian langur has a stable male dominance hierarchy which is established and maintained with a minimum of aggressive interaction. By far the majority of dominance interactions are in the form of very subtle threats, so fighting is exceedingly rare and wounds are infrequent. The male hierarchy is linear and each male is either dominant over or subordinate to each other male of his group. No coalitions or central hierarchies are formed. Since a group is normally relaxed, it is often difficult to identify the dominance structure and it may take several months of continuous observation before dominance relationships are evident to the observer.

Female langurs can be grouped into a very loosely defined and unstable hierarchy in which it is possible to delineate only general levels of dominance. It is difficult to generalize about the qualities of dominant female langurs since an individual's dominance status may be highly variable. To the extent the irritable females are more aggressive, they tend to dominate more individuals than less active and less responsive females, although there are exceptions. In general, females are more boisterous and engage in more frequent though less intense squabbles than do male langurs; but among the females, as among males, actual fighting is infrequent. The positions of many females shift continuously and markedly as the reproductive cycle progresses. A female in consort with a dominant male enjoys a temporary rise in dominance status. Similarly, a female's dominance rank tends to increase slightly during the last weeks of pregnancy. In contrast, a female with a newborn infant is effectively outside the dominance hierarchy and avoids dominance interactions whenever possible.

4. Howlers

Carpenter (1934) described the arboreal New World howler monkey as follows: "I have not observed any instances of fighting or contention between the adult males which are well integrated into a clan. The relations among clan males are best described, so far as my observations show, as being peaceful and cooperative. Howler males have not been observed to compete for sexually receptive females, for food, or for positions" (p. 99). "The social relations of the adult females . . . are mutually interactive, communal, and peaceful" (p. 78). They act in unison and rarely come into conflict. When conflicts do occur, they are resolved quickly. Carpenter noted that it is difficult to determine the status of individual females, but he judged that the females of a group can be ranked along a dominance-subordinance gradient of low slope. In general, older females are somewhat dominant over younger females.

5. APES

The sexes in the arboreal gibbon are almost indistinguishable by size, and there is no notable difference in dominance or aggressiveness between males and females (Carpenter, 1940). As a result, adults of both sexes share in leading, coordinating, and guarding the group. The gibbon appears to be the only exception to the generalization that some degree of defensive action against disturbances or predation is characteristic of adult males among monkeys and apes.

Neither male nor female chimpanzees show a clear dominance hierarchy, but the male is more robust than the female and males are always dominant over females. Reynolds and Reynolds (1965) reported that dominance interaction constituted only a minute portion of the behavior they observed, although they did note some evidence of individual status differences. There appear to be no exclusive rights to receptive estrous females and no permanent group leaders. Some adult males move with great confidence in their bearing and are relaxed. Their unhurried gait is similar to that of a very dominant langur or baboon. Quarrels among chimpanzees are rare and seldom last for more than a few seconds. Goodall (1965) reported few aggressive interactions among either males or females and also that when fights did occur they took place among animals of like sex.

Gorillas display a high degree of sexual dimorphism. Males can be ranked within a linear dominance hierarchy, every member of which can dominate all females. Females appear to lack stable hierarchies (Schaller, 1963). Insufficient data are available on orangutan behavior to comment on patterns of dominance for this great ape.

D. Reproduction

1. SEXUAL BEHAVIOR

Until the advent of systematic field studies, knowledge of sexual behavior was based on pairs of animals living in cages or in groups in artificial colonies. The limitations of these captive-animal situations are clearly reflected in distortions of behavior which were assumed for many years to represent "normal" reproduction. The cage is not a context in which one sees the reproductive *cycle* including the long periods of sexual inactivity that are normal in the field. Natural variations in day length, light, temperature, and diet importantly influence reproductive physiology and behavior (Amoroso & Marshall, 1960), but such variables are usually held constant in a laboratory and do not produce the reproductive seasonality characteristic of most species of monkeys living under normal conditions. While length of pregnancy, time of ovulation, and other basic physiological processes can best be investigated in the laboratory, the total social matrix of sexual behavior is apparent only in natural settings. Data from captive colonies where strange animals are placed together under severely overcrowded conditions have attested primarily to social disorganization. An example is the hamadryas baboon colony at the Regent's Park Zoo in London where a large proportion of animals was killed, including all but one of the newborn (Zuckerman, 1932). Added to the abnormal statistics of aggression, fatality, and sexuality was a confusion of the significance and role of behavior patterns such as dominance mounting. What appeared to be sexual behavior between emotionally disturbed members of the same sex was in fact dominance interaction.

Field observations show that constant sexual attraction is not the basis for the primate social group. For example, group structure remains stable among Japanese macaques although there is no copulation for from 6 to 8 months of the year (Mizuhara, personal communication). Sexual behavior of the female monkey or ape is regulated by her reproductive cycle, which normally includes periods of sexual receptivity followed by much longer periods of unreceptivity when sexual behavior does not occur. The normal menstrual cycle for monkeys, when a female is neither pregnant nor lactating, includes only from 5 to 14 days of sexual receptivity corresponding to the period when ovulation is likely to occur. During this time she solicits mounting by many males. When she becomes pregnant, menstrual cycles cease and the female is not usually sexually active for many months.

Lancaster and Lee (1965) refute the widely held notion that there are no seasonal differences in reproduction among primates. Studies of macaques, langurs, and baboons all indicate either a peak or a season of births in these species. The discreteness of the birth period varies geographically within a species, according to the intensity of environmental stimuli. Lancaster and Lee suggest that future investigation will show that primate species altogether lacking seasonal differences in reproduction are rare.

a. Macaques. Among free-ranging rhesus macaques in India, consort pairs last from a few hours to a few days, during which a male and estrous female move together while feeding and grooming (Fig. 12). Koford (1965) described reproductive behavior of the 400 rhesus macaques living on Cayo Santiago. Because the animals on this island are observed continuously and all animals over 1 year old are marked, the date of every birth is known within 1 day. With few exceptions, infants are born only during the first 6 months of the year. Mating occurs only during a 5- to 6-month period of the year, primarily in August, September, and October. On many males, the skin of the perineum and

FIG. 12. Adult male rhesus macaque mounts estrous adult female. (Photograph by Phyllis Jay.)

adjacent areas reddens and may become ridged a month before mating. During the nonbreeding season the testes become smaller (Sade, 1964) and spermatogenesis ceases (Conaway & Sade, 1965). Between 65% and 100% of the females reproduce annually, depending on the individual, the group, and the year. Many female rhesus macaques on Cayo Santiago display distinct estrous periods during the early part of gestation, but the adult female with a very young infant is not sexually receptive (Conaway & Koford, 1964).

In contrast to the rhesus macaque, there is very little coloration or swelling of the sexual skin among bonnet macaques observed by Simonds (1965) in southern India. Estrous females copulate with mature and inmature males of all dominance ranks throughout their period of receptivity. No consort relations were observed since the female and male usually separate after each copulation. Simonds reported that the majority of solicitations for copulation were by males and that copulation is completed in one mounting. Males examine and smell the genitalia of most females in the group each day, a pattern of behavior also observed among Ceylon toque macaques (*Macaca sinica*). I have observed that female toque macaques have a strong-smelling discharge during certain days of the menstrual cycle. Olfaction may play a more important part in sexual behavior among some species than has been suspected.

b. Baboons. The female baboon is sexually receptive only during a 12-day period of sexual swelling in a menstrual cycle of approximately 35 days. Maximum swelling lasts for about 7 days followed by a 5-day period of gradual deturgescence (DeVore, 1962). Females in Kenya baboon groups initiate sexual activity, and Bolwig (1959) recorded the same for female chacma baboons. Among groups living in the Cape in South Africa, initiation of sexual behavior depends on the female's state of swelling and on the male with her. Males sometimes initiate mounting but only with estrous females (Hall & DeVore, 1965). Olfaction may be important in baboon reproduction, since dominant males tend to copulate at the end of the 7-day period of highest sex-skin swelling (DeVore, personal communication). The number of copulations is usually highest for the most dominant males, although interruptions and harassments during copulation are common (see Section III, B). Hall and DeVore (1965) indicated that copulation is most frequent in early morning and during evening hours, which are generally the period of greatest social activity. During gestation and the early months of lactation, before her infant is 6 to 8 months old, a female baboon is behaviorally a sexless animal. In some species of baboons, a pregnant and later lactating female may not be sexually active for more than a year (DeVore, 1963b).

The composition of one-male groups of hamadryas baboons remains constant although the females in a group may not have estrous periods for months (Kummer & Kurt, 1963). Of 87 well-isolated one-male groups, the females in 72 were in the same sexual state; that is, either all or none had sexual swellings. Newborn baboons were present only in groups in which all females lacked sexual swelling. Kummer and Kurt suggested this indicates that the one-male groups may be physiologically homogeneous at least during parts of the annual cycle. Sexual activity of a male was restricted to members of his group; he did not copulate with females in another group even though his own females may not have been receptive for from 2 to 6 months.

c. *Langurs*. The female langur monkey of northern India is sexually receptive for from 5 to 7 days a month when she is not pregnant or nursing a small infant. Sexual behavior occupies an exceedingly small proportion of her adult life (Jay, 1963c, 1963d). A female langur lacks swelling and brightening of the sex skin, but she indicates her sexual receptivity by special gestures. During a consortship, which seldom lasts more than a few hours, the male follows the female but rarely threatens her (Jay, 1963a, 1963d). The female langur may solicit mounting by males for a few days in the first month of pregnancy, but for the remainder of gestation and the early months of lactation she is not sexually receptive (Jay, 1965).

d. *Apes*. Since gibbon sexual behavior occurs within the small, stable group, the young are probably progeny of the adult male in the same group. Goodall (1965) reported that persistent associations of adult male and female chimpanzees are most common in September and November, which were the only months during which she observed copulations. Schaller (1963) reported only two copulations during his observation of the mountain gorilla.

2. MATERNAL BEHAVIOR

An infant is limited in the degree to which it controls its activities and can support itself and thereby places restrictions and demands on the mother and the group. The most obvious limitations on the mother are motor and reproductive. She could move farther and faster if she did not have to carry or walk beside her infant. She could also produce more infants if the period of infant dependency were shorter.

In all species of monkeys for which data are available, it is apparent that the mother-infant relation is an important part of the group-wide matrix of social relations, and that it cannot be understood fully in isolation. The presence of infants necessitates alterations in social behavior of a group to assure the protection and development of its young members. Depending on the species there are differences in the relationship of the infant with its mother and with other adults in the group. In ground-living species, a mother with a clinging, dependent infant is less able to move quickly and to defend herself, and she remains in the most protected part of the group surrounded by adult males and females. In tree-living monkeys the mother and infant may be seen on the edge of the group as often as in its center.

Although mothers of the various species of monkeys vary considerably in the degree to which they allow other females to hold or touch infants, field data suggest that adult females of all species have a strong interest in newborn infants. The presence of young infants is a strong cohesive factor in social life and affects the daily routine as well as patterns of interaction among group members. In some terrestrial species the social bonds between adult males and infants are strong, but among tree-living species they appear to be weak or absent. This might be related to the need to respond quickly to leaders in ground-living species where survival often depends on fast responses to predators. (This is especially adaptive for juveniles, which are near the edges of the group.) These are also the species in which the dominance hierarchies are most rigidly defined and maintained.

The mother-infant relation is the most intense one in a primate group and it outlasts any other social bond. Whenever hungry, tired, or frightened the infant goes to its mother, regardless of the quality of her maternal care. Although she may be inept or rejecting or both, she remains a haven of safety and a source of emotional security (Fig. 13). The infant's close attachment to the mother creates a rich environment for social learning (see Section III, E).

The first few weeks or months of life are followed by a period of rapid growth and concomitant changes in maternal behavior. The growth in sensorimotor capabilities is accompanied by reaction against physical restraint. The infant becomes more independent, though this independence is slight at first. Physical control of the infant, holding or pulling it back, is supplemented by control based on gestural and vocal cues (Section III, B). The infant responds quickly to slight changes in the mother's posture, light touches, or soft sounds.

The early stage of dependence on the mother is followed by another in which the ties between the two are gradually loosened. The age at which weaning occurs varies among species and may depend in part on when the mother resumes menstrual cycles. Although early independence is initiated by the infant, the remaining bonds of dependency are severed by the mother during weaning, a process which appears to be primarily emotional since the infant is capable of feeding itself. The severity and rapidity of weaning varies among species, but most often rejection begins mildly and the mother usually allows the infant to sit by her side and cling when the group sleeps at night and at midday.

FIG. 13. Mother and infant Ceylon toque macaque eat together. The infant clings to his mother because he is slightly tense. Note the female's full cheek pouches. (Photograph by Phyllis Jay.)

The following survey of patterns of maternal behavior in baboons, macaques, langurs, and several of the apes indicates the range of behavior observed in the field. The wide variety of maternal behavior in free-ranging species offers a similarly wide range for experimentation. Laboratory analysis has concentrated on rhesus macaques (see Chapter 8 by Harlow and Harlow), but with the investigation of new species the components of primate maternal behavior can be investigated in greater detail. Maternal behavior depends on reflexes of both mother and infant and on behavior that the adult female has probably had to learn. The nature and importance of mother and infant behavior are clear in a field situation, but field observation alone cannot reveal what parts of these behavior patterns are learned. Similarly, it has not been possible in the laboratory to recognize the problems that normally confront mother and infant in a free-ranging condition.

a. Baboons. The normal interval between births for baboons is about 20 to 24 months and the minimum is 18 months. Since the infant of the previous pregnancy is weaned before a female baboon gives birth again, no female was observed tending two infants. Mother and infant are together most of each day for the first 6 to 8 months of an infant's life (Fig. 14). A female savanna baboon with a dependent infant could not survive for long without the protection of the social group and its large and powerful adult males. When the group moves the infant is carried either on the mother's belly or, when it is about 3 to 4 months old, on her back. If a female baboon with an infant finds it difficult to keep up with the group, adult males lag behind and walk beside her or the whole group may slow down. Having the large infant ride on her back makes it possible for a mother to keep up with the group on the daily routine of travel which averages at least 3 miles. The presence of young is one of the most important binding elements in a baboon group (DeVore, 1963b; DeVore & Washburn, 1961; Hall & DeVore, 1965). Adult females frequently approach, present to (Section III, C, 1), and then groom the mother. The mother baboon does not allow another female to touch the infant for from 7 to 10 days after its birth, but some older males occasionally approach and touch the infant and on several occasions dominant males actually carried young. The relation between infant and adult males is close at birth and grows in intensity, so that by the middle of the second year the young male baboon is more closely associated with adult males than with its own mother (DeVore, 1963b).

The dominance status of a mother baboon probably has a pronounced influence on the behavior and status of the infant. DeVore (1963b) suggested that mothers which are constantly threatened and dominated tend to be short-tempered and less responsive to their infants than are dominant mothers.

FIG. 14. Adult female baboon sits with her young infant. (Photograph by Irven DeVore.)

b. Macaques. Southwick *et al.* (1965) and Koford (1965) have discussed maternal behavior among rhesus monkeys. Since most female rhesus macaques bear an infant approximately every 12 months, mothers on Cayo Santiago are sometimes associated with as many as three offspring (Fig. 15) (Sade, personal communication). Small groups of females with their young tend to associate more closely than other combi-

FIG. 15. Adult female rhesus macaque grooms her large infant while another infant sits nearby. (Photograph by Phyllis Jay.)

nations of group members and spend several hours a day grooming and relaxing together. The mother rhesus macaque is extremely posses- sive and does not allow other females to carry her infant for the first several weeks of its life. Among rhesus monkeys living in a crowded temple area of a northern Indian city, Southwick *et al.* (1965) noted that the relation between adult males and infants or juveniles appears to be neutral or indifferent, but if a young monkey gets in the way of an adult male, particularly at feeding times, it may be severely attacked: picked up, bitten, and thrown on the ground. Examples of adult males attacking females with infants were also cited. No similar instances were reported for baboons and none was ever observed among langurs.

Bonnet macaques of southern India display maternal behavior pat- terns similar to those of the rhesus. The mother-infant relation is in- tense and continuous for the first few weeks of life. The mother bonnet macaque does not allow any other female to hold her infant for weeks after birth. Later, if a dominant female approaches and takes it, the mother holds onto the infant's arm or leg until she can take it back. Ap- proximately 5% of infants between the ages of 2 and 6 months ride on their mothers' backs occasionally but this was never observed in infants older than 6 months. Simonds (1965) noted that females do not ap- pear to play with their infants. The infant is weaned when it is between 6 and 10 months old, although when the mother gives birth again the last infant may occasionally stay near her. I observed one female bonnet macaque holding and carrying two infants younger than 2 months old; an adult member of the group had been killed by a car and one of the infants may have been that of the dead animal. The importance of this observation was that the mother did not appear to have any difficulty in holding, carrying, and nursing two infants. Both young monkeys ap- peared to be in excellent physical condition and both responded to her as other normal infants respond to their mothers.

Most female Japanese macaques bear an infant about every 12 months. Females may associate with several infants representing successive birth seasons. Paternal care among Japanese macaques at Takasakiyama, described by Itani (1959), is an important supplement to maternal be- havior. Especially during the first peak of the birth season, almost every very dominant adult male in this group protected a 1-year-old and oc- casionally a 2-year-old infant as would the mother. This is a relation between a particular male and infant during which they groom each other and the male holds and fondles the infant on his lap, walks with it, and protects it when it is threatened. Although subordinate males are usually denied access to the center of the group, a subordinate male with an infant can walk into the center of the feeding group as freely as can a female with an infant. The Japanese workers recorded that the infant assumed the status of the male that was protecting it. Approximately equal numbers of male and female yearlings are cared for by the adult males, but the majority of the 2-year-olds so cared for are females. This may be because 2-year-old males tend to form peripheral groups while young females tend to stay in the center of the group. Occasionally the same adult-young pair persists for 2 years. Most adult males associating with young monkeys become milder than normal and enter into fewer aggressive interactions, although a very few males become more aggressive. Of 18 groups of Japanese macaque, 3, including that at Takasakiyama, displayed this paternal behavior. It was rare in 7 groups, and absent in 8 (Itani, 1959).

c. *Langurs* (Fig. 16). As soon as a newborn infant is noticed, the females of a langur group crowd around the mother and all try to touch, hold, and carry the infant. Within a few hours of birth the mother allows a waiting female to take the infant (Jay, 1962). For several hours of the first day of life a newborn may be held by most of the adult and sub-adult females in the group and be carried as far as 50 feet from the

Fig. 16. Mother langur holds her newborn infant. It has just dried and she has not yet allowed another female to hold it. (Photograph by Phyllis Jay.)

mother. The mother langur carries her infant during periods of stress (Fig. 17), but when the group is relaxed, as it is most of the time, she allows other females to take the infant for hours every day. Unlike baboons, mother langurs do not orient toward dominant males, and usually tend to avoid them when their infants are very young. Adult male northern Indian langurs evince no interest in infants. The female usually

FIG. 17. The newborn infant northern Indian langur can cling to its mother within an hour after birth, even when she runs rapidly. (Photograph by Phyllis Jay.)

completely disassociates herself from the previous infant before she gives birth again. Ripley (personal communication) reports similar behavior for the Ceylon gray langur (*Presbytis entellus thersites*).

Considerable variability in maternal behavior is observed when the observer recognizes many females individually. Although most mother langurs are extremely competent (Fig. 18), a few are less so and have occasional difficulty in keeping an infant quiet and contented. Only one female was observed to have great difficulty holding a young infant, making it cling, and keeping it quiet. More irritable, easily disturbed females tend to move suddenly and often startle their own and nearby infants (Jay, 1963a, 1963b). A few females weaned their young completely within the first few weeks, but most mother langurs prolonged the weaning process for 3 to 5 months. At the end of this time it was no longer possible to determine which young juveniles were the offspring of which adult females.

When the group is disturbed and moves suddenly, a mother langur's first concern is normally the protection of her infant. But once, when a group of southern Indian langurs was being chased by man, a mother pushed aside a squealing infant about 2.5 months old and the infant fell almost 20 feet to a lower limb of the tree. It clung tenaciously with one foot and squealed loudly. The mother jumped into the next tree and the infant hung screaming for almost 20 seconds until another adult (unidentified) dashed down to it and grabbed it with one hand to take it quickly to the top of the tree. Similar but equally rare instances of a mother pushing aside an infant when the group is suddenly upset were reported by Stott (personal communication) for the African black-and-white colobus monkey. For several weeks after birth, the infant olive colobus is carried in the mother's mouth when she moves. After that time it clings to the mother's belly (Booth, 1957).

d. Apes. Infant dependency among apes is substantially longer than that of any monkey. Lengthening the period of infant dependency undoubtedly has important effects since it extends the initial learning stage when the infant is assured of care and protection by the adults of the group.

The varying relation between infant capabilities and maternal behavior is observed when the relatively helpless young chimapanzee is compared to the precocious infant macaque. The chimpanzee has developed far more elaborate maternal behavior patterns than the macaque or any other monkey. Goodall (1965) reported that mother chimpanzees carry infants for 2.5 to 3 years. The mother-infant bonds in the groups she studied probably persisted into adult life, and she suggests that they may form the main bond in a social group composed of fe-

Fig. 18. Adult female southern Indian langur (*Presbytis entellus priamus*) plays with her 1-month-old infant. (Photograph by Phyllis Jay.)

males with infants, older juveniles, adolescents, and young mature animals. Weaning as an emotional rejection, characteristic of monkeys, does not appear among chimpanzees. Gradual independence is initiated by the young and no behavior fitting the traditional description of weaning was observed. On several occasions, Reynolds and Reynolds (1965) saw mother chimpanzees carrying an infant on the belly at the same time a juvenile rode on the back. In each instance the young rode without active support by the mother.

The mother gorilla must support her infant with one arm for the first 3 months; then it begins to ride on her back (Schaller, 1963). The mother-infant bonds are close for the first 3 years of the infant's life, after which the young gorilla is usually independent although it may have some association with its mother for another 1.5 years.

E. Socialization

The general features of socialization are similar for most species of monkeys and apes, although variations do occur. The infant primate learns to apply behavior patterns characteristic of the species as well as a great many patterns specific to its particular group and to its environment. Repetitiveness and routine characterize the world of young primates and provide an extremely stable surrounding in which change is not one of the problems facing the individual.

It is well documented that the behavior of adult humans is influenced by what happened during infancy, and it is becoming more apparent that this also holds true for nonhuman primates. With the retardation of growth, and longer infant dependency, there is a tendency for individual experience to play a more subtle and detailed role in shaping instinctive behavior into biologically effective patterns (Mason, 1965). The very young infant with an immature sensory system is probably an immature learner (see Chapter 10 by Fantz and Chapter 11 by Zimmermann and Torrey), but it quickly turns into a rapid learner. The extended period of infant dependency in primates may increase the amount and complexity of learning that is possible. This increases the opportunity to shape the behavior of the young to meet local environmental conditions. Flexibility of patterns of behavior may be one of the principal gains from a long learning period. Mason (1965) has suggested that learning is more important in the development of social behavior than has been thought. For example, exploration, sex, fighting, dominance, nest building, and other elements of behavior are in part learned during play activity. But the relative role of inherited and environmental factors may be very different for different items of behavior.

Social attachments (responsiveness to other members of the species) are probably formed early in life. The attractiveness of young animals to adult females and in some instances to adult males usually insures constant attention from many members of the group and attachments are formed easily. Any extension of the period of infant dependency would necessarily increase the length of time during which these initial attachments are formed.

Infant responses which appear first are those which help the infant to maintain bodily contact with the mother and to feed (see Chapter 8 by Harlow and Harlow). Many of these components of infant behavior have been analyzed in the laboratory. For example, a high level of arousal strengthens the tendency to cling (see Chapter 9 by Mason). This is certainly apparent in the field. Clinging, in turn, probably reduces the level of infant arousal or distress and may constitute a major factor in the development of attachment to the mother.

As the infant matures, patterns of behavior developed in the mother-infant relation are strengthened and broadened to include other animals. The social world of the infant rapidly expands to include more of the group, including age-mates. Many social contacts are made on the initiative of the infant. Infants of some species do not contact certain classes of adults until they are older, but infants of other species may become acquainted with most group members in the first months of relative independence. The greatest variable in infant contact appears to be with adult males, since in some kinds of monkey, such as langurs, the young have little or no contact with adult males. Because social groups in most species of primates are stable in membership, the young animal soon learns which members to avoid and which others are friendly or indifferent.

When infants contact age-mates and start to move about independently of the mother, play becomes an extremely important daily activity. The 6-month-old infant may play with young monkeys for 3 to 5 hours a day. The general descriptive term "play" includes a wide variety of activities, and although it is difficult to define the components of play there is great agreement among observers in judging "playful behavior" in a field situation. During these early years, adult members of the group are very tolerant of the young. At least among langurs and baboons, accidental contacts with adult males do not result in threats or injury. This high degree of early tolerance is gradually withdrawn as the young mature, and finally the young learn which responses to threat will avoid aggressive repercussions and which threats may safely be directed to other monkeys. Effective use of gestures and vocalizations can prevent serious harm to an individual. Although the adult repertoire is built primarily by addition, several gestures and calls are dropped as the young monkey grows older. Before sexual maturity, in most species for which data are available the juvenile is familiar with all patterns of social interaction in the group.

Among all species for which data are available, gestures and vocalizations of aggressive dominance first appear in a context of maximum safety for the young animal (DeVore, 1963a; Jay, 1965). This context is usually the play group composed of the like-aged and the like-sized, where the young monkey can practice and is not likely to be hurt if it threatens age mates. Playing monkeys may run from one end of the group to the other but they are always near adults whose watchfulness prevents danger from predators. If play becomes too rough, it is stopped by adults; but playful wrestling and fighting are the youthful opportunities for acquiring skillful use of aggressive behavior patterns.

The age of transition from social immaturity to adulthood differs between male and female monkeys. Female baboons and langurs are socially mature when they give birth (usually a comparatively short time after puberty), which may be 3 to 5 years before a male of the same age achieves his social maturity (DeVore, 1962; Jay, 1963a). The male achieves social maturity when he attains a status position in the dominance structure which cannot happen until he is fully mature physically (i.e., has fully erupted canines and the muscles and body size of an adult male). Although actual fighting among males is infrequent, the male's status is ultimately based on his ability to defend it and this requires the social learning and the final physical growth of the years following puberty.

Intensive, detailed analysis of socialization will have to be undertaken in future field studies of longer duration than those previously completed. The study of the northern Indian langur is one of the few recent studies which has concentrated on aspects of socialization. The Japanese macaque has been studied by the Japan Monkey Centre for a longer time than any other species to date, and the complete records of known animals will be invaluable in understanding many patterns of socialization in this species.

Laboratory experimentation has demonstrated that social deprivation has profound effects on the maturing animal (see Chapter 8 by Harlow and Harlow). Most laboratory research on social deprivation has concerned gross variations in rearing conditions. Physical growth of primates does not appear to be directly affected by social factors, but many aspects of psychological development depend on social experience (Mason, 1965). The mother-infant relation is apparently less important than infant-infant relations in normal social development of caged rhesus monkeys (Harlow, 1958, 1959, 1962; Harlow & Harlow, 1962). Development was essentially normal, according to the standards of normality defined by the experimenters, if the individuals were raised by their mothers and given daily opportunity to interact with age-mates. Those infants deprived of their mothers but given frequent contact with peers were initially retarded in social development but gradually developed "normal" patterns. Infants in contact only with their mothers were most retarded. One of the most severe deficiencies in socially deprived monkeys is inadequate sexual behavior. The various constituents of the male mating pattern probably do not develop as a unit but appear at different stages in ontogeny and are differentially related to experience. Age-mates are available to the young monkey and to the young gorilla and chimpanzee in the normal group structure of the species. The gibbon presents unique adaptations in infant development since the social group is normally composed of only one adult male, one adult female, and her young. Infant gibbons lack age-mates but may play with siblings if there are any.

An understanding of the complexities of primate socialization will result only from continuous correlation and checking of field data with laboratory data and the planning of investigations based on both. The total web of social learning, play and exploration, sensorimotor coordinations, and reflex systems can only be described tentatively, at best, from observation of free-ranging primates. If the complex factors which determine the behavior of mature animals living in a social group are to be discovered and evaluated, detailed experimental analysis is essential.

IV. PERSPECTIVES

Social behavior of nonhuman primates living under natural conditions is complex and richly varied. This chapter has surveyed only a few of many possible topics of primate behavior and has done so for only a few of the species of living monkeys and apes. Primate behavior has been investigated in laboratories, artificial colonies, and natural free-ranging situations, and the research within each of these three contexts has produced different kinds of knowledge and insight into the behavior of monkeys and apes. Data resulting from these different approaches are mutually supplemental, and all the approaches are essential in the analysis of primate behavior.

Since biological adaptation can be defined as repeatedly successful behavior, the usefulness of field studies is their concern with the observation of behavior as it occurs in its natural environment. Comparative study reveals behavioral differences which assume their proper significance as adaptations tending to maximize the likelihood of species survival in specific habitats. For example, highly organized and rigidly structured groups led by large, powerful adult males can survive on an open savanna where predators would soon destroy monkeys whose immediate response to danger is to scatter and run for trees. Anatomical and social specializations advantageous in coping with danger have developed among those species whose wide range of diet allowed them to invade open treeless areas.

Although the adaptive significance of complex behavior patterns can be understood only by viewing them over time in a natural context, their anatomical and physiological components can be analyzed in detail only under laboratory conditions where variables are carefully controlled.

Our knowledge of basic patterns of behavior among nonhuman primates has great relevance for deeper understanding of man's own adaptations, but this is beyond the scope of the present chapter. Man has survived during the course of evolution as a very effective primate. Much of his present behavior can be understood better against the background of his primate biological and social heritage. Although his culture has largely freed him from the physical demands of his environment, his emotions and physiology are best analyzed as aspects of complex patterns of adaptation which have enabled him to survive through various stages of social evolution. Knowledge derived from modern field studies of monkeys and apes has much to contribute to the study of man himself, the most successful, complex, and widely distributed of the living primates.

ACKNOWLEDGMENTS

This chapter would not have been possible without the generous cooperation of all the members of the Primate Project held at the Center for Advanced Study in the Behavioral Sciences, Stanford, 1962–1963. Many of the statements and generalizations made here are based on the results of recent and as yet unpublished field observations. Contributions to the book on primate behavior (DeVore, 1965) written by the participants in the primate project were made available for this chapter, as were additional unpublished material and field notes.

Irven DeVore, K. R. L. Hall, David A. Hamburg, Hiroki Mizuhara, Vernon and Frances Reynolds, George B. Schaller, Sherwood L. Washburn, and I were continuing members of the project. Other contributors to the volume include Jarvis R. Bastian, C. R. Carpenter, Jane Goodall, Carl B. Koford, Jane B. Lancaster, Richard B. Lee, Peter Marler, William A. Mason, Paul Simonds, and Charles H. Southwick. Harry F. Harlow, Stuart A. Altmann, John Kaufmann, and Arthur J. Riopelle contributed to the major conferences held to plan the volume. I express my sincere thanks and appreciation for valuable and constructive comments made on this chapter by Sherwood L. Washburn, George B. Schaller, Irven DeVore, Jane B. Lancaster, Chesley Lancaster, Vernon and Frances Reynolds, K. R. L. Hall, and Anne Brower. The writing of this chapter was undertaken during, and made possible by, a Fellowship at the Center for Advanced Study in the Behavioral Sciences, and the resulting extended opportunity for collaboration with the other Fellows at the Center. The primate project was supported in part by a grant (M-5502) from the U. S. Public Health Service.

Many of the photographs included in this chapter were contributed by members of the primate project. A large number of the photographs contributed by me were taken while I was a technical advisor to a photographer from *Life* magazine during a revisit to India, Ceylon, Southeast Asia, and Japan in February and March of 1963. I gratefully acknowledge the support of *Life* magazine during this trip.

Interpretations and generalizations in this chapter must of necessity be tentative since many species of primates have not been observed. Stimulation for many interpretations has originated among my colleagues, and I have attempted to indicate in the text where these occur. However, the fact that the participants in the conferences and writers of the volume have contributed their data so unstintingly does not commit them to support of generalizations included in this chapter.

REFERENCES

Altmann, S. A. (1960). A field study of the sociobiology of rhesus monkeys, *Macaca mulatta*. Unpublished doctoral dissertation, Harvard University.

Altmann, S. A. (1962). A field study of the sociobiology of rhesus monkeys, *Macaca mulatta*. *Ann. N. Y. Acad. Sci.* **102**, 338.

Amoroso, E. C., & Marshall, F. H. A. (1960). External factors in sexual periodicity. *In* "Marshall's Physiology of Reproduction" (A. S. Parkes, ed.), 3rd ed., Vol. I, Part 2, pp. 707–831. Longmans, London.

Andrew, R. J. (1962). Evolution of intelligence and vocal mimicking. *Science* **137**, 585.

Andrew, R. J. (1963). The origin and evolution of the calls and facial expressions of the primates. *Behaviour* **20**, 1.

Beatty, H. (1951). A note on the behavior of the chimpanzee. *J. Mammal.* **32**, 118.

Bernstein, I. S. (1962). Response to nesting materials of wild born and captive born chimpanzees. *Anim. Behav.* **10**, 1.

Bierens de Haan, J. A. (1931). Werkzeuggebrauch und Werkzeugherstellung bei einem niederen Affen. (*Cebus hypoleucus* Humb.). *Z. vergl. Physiol.* **13**, 639.

Bolwig, N. (1959). A study of the behaviour of the chacma baboon, *Papio ursinus*. *Behaviour* **14**, 136.

Booth, A. H. (1955). Speciation in the mona monkeys. *J. Mammal.* **36**, 434.

Booth, A. H. (1956). The distribution of primates in the Gold Coast. *J. W. Afr. Sci. Ass.* **2**, 122.

Booth, A. H. (1957). Observations on the natural history of the olive colobus monkey, *Procolobus verus* (van Beneden). *Proc. zool. Soc. Lond.* **129**, 421.

Booth, A. H. (1958). The Niger, the Volta and the Dahomey Gap as geographic barriers. *Evolution* **12**, 48.

Bourlière, F. (1956). "The Natural History of Mammals," 2nd ed. Knopf, New York.

Burt, W. H. (1943). Territoriality and home range concepts as applied to mammals. *J. Mammal.* **24**, 346.

Burt, W. H. (1949). Territoriality. *J. Mammal.* **30**, 25.

Carpenter, C. R. (1934). A field study of the behavior and social relations of howling monkeys (*Alouatta palliata*). *Comp. Psychol. Monogr.* **10**, No. 2 (Whole No. 48).

Carpenter, C. R. (1935). Behavior of red spider monkeys in Panama. *J. Mammal.* **16**, 171.

Carpenter, C. R. (1940). A field study in Siam of the behavior and social relations of the gibbon (*Hylobates lar*). *Comp. Psychol. Monogr.* **16**, No. 5 (Whole No. 84).

Carpenter, C. R. (1942). Sexual behavior of free-ranging rhesus monkeys (*Macaca mulatta*). *J. comp. Psychol.* **33**, 113.

Carpenter, C. R. (1958). Territoriality: a review of concepts and problems. *In* "Behavior and Evolution" (Anne Roe & G. G. Simpson, eds.), pp. 224–250. Yale Univer. Press, New Haven, Connecticut.

Carpenter, C. R. (1965). The howlers of Barro Colorado Island. *In* "Primate Behavior: Field Studies of Monkeys and Apes" (I. DeVore, ed.), pp. 250–291. Holt, Rinehart & Winston, New York.

Cherry, C. (1957). "On Human Communication." Technol. Press of Mass. Inst. of Technol., Cambridge, Massachusetts, and Wiley, New York.

Collias, N., & Southwick, C. (1952). A field study of population density and social organization in howling monkeys. *Proc. Amer. phil. Soc.* **96**, 143.

Conaway, C. H., & Koford, C. B. (1964). Estrous cycles and mating behavior in a free-ranging band of rhesus monkeys. *J. Mammal.* **45**, 577.

Conaway, C. H., & Sade, D. S. (1965). The seasonal spermatogenic cycle in free ranging rhesus monkeys. *Folia primatol.* **3**, 1.

Dasmann, F. R., & Mossman, A. S. (1962). Population studies of impala in Southern Rhodesia. *J. Mammal.* **43**, 375.

DeVore, B. I., Jr. (1962). Social behavior and organization of baboon troops. Unpublished doctoral dissertation, University of Chicago.

DeVore, I. (1963a). A comparison of the ecology and behavior of monkeys and apes. *In* "Classification and Human Evolution" (S. L. Washburn, ed.), pp. 301–319. Wenner-Gren Foundation for Anthropological Research, New York.

DeVore, I. (1963b). Mother-infant relations in free-ranging baboons. *In* "Maternal Behavior in Mammals" (Harriet L. Rheingold, ed.), pp. 305–335. Wiley, New York.

DeVore, I. (ed.) (1965). "Primate Behavior: Field Studies of Monkeys and Apes." Holt, Rinehart & Winston, New York.

DeVore, I., & Hall, K. R. L. (1965). Baboon ecology. *In* "Primate Behavior: Field Studies of Monkeys and Apes" (I. DeVore, ed.), pp. 20–52. Holt, Rinehart & Winston, New York.

DeVore, I., & Lee, R. (1963). Recent and current field studies of primates. *Folia Primatol.* **1**, 66.

DeVore, I., & Washburn, S. L. (1961). Baboon behavior. Univer. Extension, Univer. Calif. Berkeley. (Film)

DeVore, I., & Washburn, S. L. (1963). Baboon ecology and human evolution. *In* "African Ecology and Human Evolution" (F. C. Howell & F. Bourlière, eds.), pp. 335–367. Wenner-Gren Foundation for Anthropological Research, New York.

Furuya, Y. (1960). An example of fission of a natural troop of Japanese monkey at Gagyusan. *Primates: J. Primatol.* **2**, 149.

Goodall, Jane M. (1962). Nest building behavior in the free-ranging chimpanzee. *Ann. N. Y. Acad. Sci.* **102**, 455.

Goodall, Jane (1963). Feeding behaviour of wild chimpanzees. *Sympos. zool. Soc. Lond.* **10**, 39.

Goodall, Jane (1965). Chimpanzees of the Gombe Stream Reserve. *In* "Primate Behavior: Field Studies of Monkeys and Apes" (I. DeVore, ed.), pp. 425–473. Holt, Rinehart & Winston, New York.

Haddow, A. J. (1952). Field and laboratory studies on an African monkey, *Cercopithecus ascanius schmidti* Matschie. *Proc. zool. Soc. Lond.* **122**, 297.

Hall, K. R. L. (1962). The sexual, agonistic and derived social behaviour patterns of the wild chacma baboon, *Papio ursinus. Proc. zool. Soc. Lond.* **139**, 283.

Hall, K. R. L. (1963a). Variations in the ecology of the chacma baboon, *Papio ursinus. Sympos. zool. Soc. Lond.* **10**, 1.

Hall, K. R. L. (1963b). Tool using performances as indicators of behavioral adaptability. *Current Anthropol.* **4**, 479.

Hall, K. R. L., & DeVore, I. (1965). Baboon social behavior. *In* "Primate Behavior: Field Studies of Monkeys and Apes" (I. DeVore, ed.), pp. 53–110. Holt, Rinehart & Winston, New York.

Harlow, H. F. (1958). The nature of love. *Amer. Psychologist* **13**, 673.

Harlow, H. F. (1959). Love in infant monkeys. *Sci. Amer.* **200** (6), 68.

Harlow, H. F. (1962). The development of affectional patterns in infant monkeys. *In* "Determinants of Infant Behavior" (B. M. Foss, ed.), pp. 75–88. Wiley, New York.

Harlow, H. F., & Harlow, Margaret K. (1962). Social deprivation in monkeys. *Sci. Amer.* **207** (5), 136.

Hinde, R. A., & Rowell, T. E. (1962). Communication by postures and facial expressions in the rhesus monkey (*Macaca mulatta*). *Proc. zool. Soc. Lond.* **138**, 1.

Imanishi, K. (1957). Identification: a process of enculturation in the subhuman society of *Macaca fuscata. Primates: J. Primatol.* **1**, 1.

Imanishi, K. (1960). Social organization of subhuman primates in their natural habitat. *Current Anthropol.* **1**, 393.

Itani, J. (1954). Japanese monkeys in Takasakiyama. *In* "Social Life of Animals in Japan" (K. Imanishi, ed.), Vol. 2. Kobunsha, Tokyo. (Partial translation by J. Frisch, mimeograph.)

Itani, J. (1959). Paternal care in the wild Japanese monkey, *Macaca fuscata fuscata. Primates: J. Primatol.* **2**, 61.

James, W. W. (1960). "Jaws and Teeth of Primates." Pitman Med. Publications, London.

Jay, Phyllis C. (1962). Aspects of maternal behavior among langurs. *Ann. N. Y. Acad. Sci.* **102**, 468.

Jay, Phyllis C. (1963a). The ecology and social behavior of the Indian langur monkey. Unpublished doctoral dissertation, University of Chicago.

Jay, Phyllis (1963b). The Indian langur. *In* "Primate Social Behavior" (C. H. Southwick, ed.), pp. 114–123. Van Nostrand, Princeton, New Jersey.

Jay, Phyllis (1963c). Mother-infant relations in langurs. *In* "Maternal Behavior in Mammals" (Harriet L. Rheingold, ed.), pp. 282–304. Wiley, New York.

Jay, Phyllis C. (1963d). The female primate. *In* "Man and Civilization: The Potential of Woman" (S. M. Farber & R. H. L. Wilson, eds.), pp. 3–12. McGraw-Hill, New York.

Jay, Phyllis (1965). The common langur of north India. *In* "Primate Behavior: Field Studies of Monkeys and Apes" (I. DeVore, ed.), pp. 197–249. Holt, Rinehart & Winston, New York.

Kaufmann, J. H. (1962). Ecology and social behavior of the coati, *Nasua narica*, on Barro Colorado Island, Panama. *Univer. Calif. Publ. Zool.* **60**, 95.

Kawai, M. (1958). On the rank system in a natural group of Japanese monkeys. *Primates: J. Primatol.* **1**, 111.

Kawai, M. (1960). A field experiment on the process of group formation in the Japanese monkey (*Macaca fuscata*), and the releasing of the group at Ohirayama. *Primates: J. Primatol.* **2**, 181.

Kawamura, S. (1958). The matriarchal social order in the Minoo-B group. *Primates: J. Primatol.* 1, 149.

Kawamura, S. (1959). The process of sub-culture propagation among Japanese macaques. *Primates: J. Primatol.* 2, 43.

Klüver, H. (1933). "Behavior Mechanisms in Monkeys." Univer. Chicago Press, Chicago, Illinois.

Klüver, H. (1937). Re-examination of implement-using behavior in a cebus monkey after an interval of three years. *Acta psychol.* 2, 347.

Koford, C. B. (1965). Population dynamics of rhesus monkeys on Cayo Santiago. *In* "Primate Behavior: Field Studies of Monkeys and Apes" (I. DeVore, ed.), pp. 160–174. Holt, Rinehart & Winston, New York.

Kortlandt, A. (1962). Chimpanzees in the wild. *Sci. Amer.* 206 (5), 128.

Kummer, H., & Kurt, F. (1963). Social units of a free-living population of hamadryas baboons. *Folia primatol.* 1, 4.

Lancaster, Jane B., & Lee, R. B. (1965). The annual reproductive cycle in monkeys and apes. *In* "Primate Behavior: Field Studies of Monkeys and Apes" (I. DeVore, ed.), pp. 486–513. Holt, Rinehart & Winston, New York.

Marler, P. (1961). The logical analysis of animal communication. *J. theor. Biol.* 1, 295.

Marler, P. (1965). Communication in monkeys and apes. *In* "Primate Behavior: Field Studies of Monkeys and Apes" (I. DeVore, ed.), pp. 544–584. Holt, Rinehart & Winston, New York.

Mason, W. A. (1965). The social development of monkeys and apes. *In* "Primate Behavior: Field Studies of Monkeys and Apes" (I. DeVore, ed.), pp. 514–543. Holt, Rinehart & Winston, New York.

Merfield, F. G., & Miller H. (1956). "Gorilla Hunter." Farrar, Straus & Cudahy, New York.

Napier, J. (1962). Monkeys and their habitats. *New Scientist* 15, 88.

Phillips, W. W. A. (1935). "Manual of the Mammals of Ceylon." Dulan, London.

Pitelka, F. A. (1959). Numbers, breeding schedule, and territoriality in pectoral sandpipers in northern Alaska. *Condor* 61, 233.

Pitman, C. R. S. (1931). "A Game Warden among his Charges." Nisbet, London.

Reynolds, V., & Reynolds, Frances (1965). Chimpanzees of the Budongo Forest. *In* "Primate Behavior: Field Studies of Monkeys and Apes" (I. DeVore, ed.), pp. 368–424. Holt, Rinehart & Winston, New York.

Rowell, T. E. (1962). Agonistic noises of the rhesus monkey (*Macaca mulatta*). *Sympos. zool. Soc. Lond.* 8, 91.

Rowell, T. E., & Hinde, R. A. (1962). Vocal communication by the rhesus monkey (*Macaca mulatta*). *Proc. zool. Soc. Lond.* 138, 279.

Sade, D. S. (1964). Seasonal cycle in size of testes of free-ranging *Macaca mulatta*. *Folia primatol.* 2, 171.

Sanderson, I. T. (1957). "The Monkey Kingdom." Hanover House, Garden City, New York.

Schaller, G. B. (1961). The orang-utan in Sarawak. *Zoologica* 46, 73.

Schaller, G. B. (1963). "The Mountain Gorilla: Ecology and Behavior." Univer. Chicago Press, Chicago, Illinois.

Schaller, G. B. (1965a). Field procedures. *In* "Primate Behavior: Field Studies of Monkeys and Apes" (I. DeVore, ed.), pp. 623–629. Holt, Rinehart & Winston, New York.

Schaller, G. B. (1965b). The behavior of the mountain gorilla. *In* "Primate Behavior: Field Studies of Monkeys and Apes" (I. DeVore, ed.), pp. 324–367. Holt, Rinehart & Winston, New York.

Schneirla, T. C. (1950). The relationship between observation and experimentation in the field study of behavior. *Ann. N. Y. Acad. Sci.* **51**, 1022.

Schultz, A. H. (1956). The occurrence and frequency of pathological and teratological conditions and of twinning among non-human primates. *In* "Primatologia" (H. Hofer, A. H. Schultz, & D. Starck, eds.), Vol. I, pp. 965-1014. Karger, New York.

Simonds, P. E. (1965). The bonnet macaque in south India. *In* "Primate Behavior: Field Studies of Monkeys and Apes" (I. DeVore, ed.), pp. 175–196. Holt, Rinehart & Winston, New York.

Southwick, C. H., Beg, M. A., & Siddiqi, M. R. (1961a). A population survey of rhesus monkeys in villages, towns and temples of northern India. *Ecology* **42**, 538.

Southwick, C. H., Beg, M. A., & Siddiqi, M. R. (1961b). A population survey of rhesus monkeys in northern India: II. Transportation routes and forest area *Ecology* **42**, 698.

Southwick, C. H., Beg, M. A., & Siddiqi, M. R. (1965). Rhesus monkeys in north India. *In* "Primate Behavior Field Studies of Monkeys and Apes" (I. DeVore, ed.), pp. 111–159. Holt, Rinehart & Winston, New York.

Starck, D., & Frick, H. (1958). Beobachtungen an äthiopischen Primaten. *Zool. Jb.* (Syst.) **86**, 41.

Sugiyama, Y. (1960). On a division of a natural troop of Japanese monkeys at Takasakiyama. *Primates: J. Primatol.* **2**, 109.

Tappen, N. C. (1960). Problems of distribution and adaptation of the African monkeys. *Current Anthropol.* **1**, 91.

Ullrich, W. (1961). Zur Biologie und Soziologie der Colobusaffen (*Colobus guereza caudata* Thomas 1885). *Zool. Garten* **25**, 305.

van Hooff, J. A. R. A. M. (1962). Facial expressions in higher primates. *Sympos. zool. Soc. Lond.* **8**, 97.

Washburn, S. L. (1957). Ischial callosities as sleeping adaptations. *Amer. J. phys. Anthropol.* **15**, 269.

Washburn, S. L., & Avis, Virginia (1958). Evolution of human behavior. *In* "Behavior and Evolution." (Anne Roe & G. G. Simpson, eds.), pp. 421–436. Yale Univer. Press, New Haven, Connecticut.

Washburn, S. L., & DeVore, I. (1961). The social life of baboons. *Sci. Amer.* **204** (6), 62.

Washburn, S. L., & Hamburg, D. A. (1965a). The study of primate behavior. *In* "Primate Behavior: Field Studies of Monkeys and Apes" (I. DeVore, ed.), pp. 1–13. Holt, Rinehart & Winston, New York.

Washburn, S. L., & Hamburg, D. A. (1965b). The implications of primate research. *In* "Primate Behavior: Field Studies of Monkeys and Apes" (I. DeVore, ed.), pp. 607–622. Holt, Rinehart & Winston, New York.

Yerkes, R. M., & Yerkes, Ada W. (1929). "The Great Apes." Yale Univer. Press, New Haven, Connecticut.

Zuckerman, S. (1932). "The Social Life of Monkeys and Apes." Kegan Paul, London.

Appendix

As the chapters of this book amply demonstrate, not all primates be-
have alike. Different species, and even different subspecies, may differ
in behavior. Although this is widely recognized, an important implication for behavioral research is often ignored. Because the results of a
study may depend on the particular primate being studied, it is impera-
tive that the researcher identify his subjects as fully and accurately as
possible.

All too often, only a vernacular name is mentioned in the report of a
behavioral investigation. But a vernacular name is not adequate identification; usage varies from time to time. This variation is reflected in a
number of differences between vernacular names used in this book and
those used when the studies being described were originally published.

In this book, the first time that a particular vernacular name appears
in each chapter, the scientific name of the animal is also given. Often
only the genus can be named; this further reflects inadequate identification in the original publication.

Even when a Latin name was given in the original study, the same
subjects may be identified with a different Latin name in this book. Such
differences sometimes arise from changes in accepted nomenclature, but
perhaps more often they result from the marked differences among
primate taxonomists in their systems of classification and nomenclature.
Any exhortation to use scientific names in identifying primates must
be tempered by the knowledge that there is no single authoritative source
for primate nomenclature. Fortunately, cases in which the same name is
used for different species are now quite rare, but the literature often contains different names for the same species.

To avoid such inconsistencies within the covers of this book, the scientific names used are those given by Fiedler (1956). Fiedler's nomenclature was adopted because it is recent and comprehensive, and also because it conforms in general to the scientific nomenclature familiar to
behavioral scientists. An example of a recent and comprehensive system
that differs markedly from familiar usage is that of Sanderson (1957), in
which, for example, the macaques (usually considered a single genus,
Macaca) are classified as a subfamily and split into eight genera.

For the benefit of readers who would like to see how various genera
are related to the rest of the Order Primates, or who want an overview

593

Appendix

TABLE I
SOME PROBLEMS OF PRIMATE NOMENCLATURE

Scientific name (Fiedler, 1956)	Vernacular name(s)	Other Latin names sometimes used
Callithrix	Marmoset	*Hapale*
Saimiri sciureus	Squirrel monkey	*Saimiri sciurea*
Cebus albifrons	White-fronted capuchin Cinnamon ringtail	
Cebus apella	Hooded capuchin Large-headed capuchin Tufted capuchin	*Cebus fatuellus* (for hooded capuchin)
Macaca irus[a]	Cynomolgus macaque Crab-eating macaque Java monkey	*Macaca fascicularis* *Macaca mulatta fascicularis*
Papio comatus	Chacma baboon	*Papio ursinus* *Papio porcarius*
Papio cynocephalus	Yellow baboon Sphinx baboon	*Papio sphinx*[b]
Papio doguera	Anubis baboon Doguera baboon Olive baboon[c] Guinea baboon[c]	*Papio anubis*
Papio leucophaeus	Drill	*Mandrillus leucophaeus*
Papio sphinx	Mandrill	*Mandrillus sphinx*
Cercocebus torquatus atys	Sooty mangabey	*Cercocebus fuliginosus*
Cercopithecus aethiops	Green monkey Vervet Grivet	*Cercopithecus sabaeus* *Cercopithecus pygerythrus*
Presbytis entellus	Indian langur Hanuman langur Sacred langur	*Semnopithecus entellus*
Presbytis johni	Nilgiri langur	*Kasi johni*
Presbytis senex	Purple-faced langur	*Kasi vetulus*
Pan troglodytes	Chimpanzee	*Pan satyrus*

[a]Fooden (1964) claims that *"irus"* is not a valid name and should be replaced by *"fascicularis."* He also claims that the cynomolgus macaque and rhesus macaque are conspecific. If both of these claims are true, the cynomolgus macaque is *Macaca mulatta fascicularis* and the rhesus macaque is *Macaca mulatta mulatta*. There have been other recent indications that all macaques belong to a single species, and there have also been hints that baboons and macaques should be classified in the same genus.

[b]This is the name used by Fiedler (1956) for the mandrill.

[c]*Papio papio*, not listed in this table, is also called "olive baboon" and "Guinea baboon" as well as "western baboon."

of the whole order, a classification of the living primates is presented on the endpapers of each volume. The classification is essentially that of Fiedler (1956); the sequence is changed slightly to list the Callithricidae before the Cebidae (and the Callimiconinae before the Cebinae), and Fiedler's Infraorders Tupaiiformes, Lemuriformes, Lorisiformes, and Tarsiiformes are listed as the Superfamilies Tupaioidea, Lemuroidea, Lorisoidea, and Tarsioidea. Probably none of these admittedly arbitrary changes would seriously bother even a taxonomist; most of the changes, in fact, are in accord with classifications by Simpson (1945) and by Straus (1949), which are in other respects similar to Fiedler's.

The use of a consistent nomenclature is necessary, but it doesn't eliminate alternative names from the literature. Some readers may be familiar with names from other nomenclatures, and they may fail to recognize a familiar species discussed under an unfamiliar name. This is one reason for using vernacular names throughout this book in addition to scientific names. A few possible sources of confusion are listed in Table I.

REFERENCES

Fiedler, W. (1956). Übersicht über das System der Primates. In "Primatologia" (H. Hofer, A. H. Schultz, & D. Starck, eds.), Vol. I, pp. 1–266. Karger, New York.
Fooden, J. (1964). Rhesus and crab-eating macaques: intergradation in Thailand. Science 143, 363.
Sanderson, I. T. (1957). "The Monkey Kingdom." Hanover House, Garden City, New York.
Simpson, G. G. (1945). The principles of classification and a classification of mammals. Bull. Amer. Mus. nat. Hist. 85, 1.
Straus, W. L., Jr. (1949). The riddle of man's ancestry. Quart. Rev. Biol. 24, 200.

Author Index

Numbers in italic indicate the pages on which the complete references are listed.

1

Subject Index

Food preferences—*Continued*
    rhesus macaque, 11, 34–35, 507
    social transmission, 538–539
Foodwells, 7
    distance between, 141–142, 197
Forced-choice procedure in oddity learn-
        ing, 195–197
Form-concept formation, 199
Form discrimination
    chimpanzee, 4–7, 372–373, 398, 456
    comparative behavior, 259
    contour regularity and, 17–18
    human infant, 377–378
    learning set, 40–41
    rhesus macaque, 13–20, 32–33, 37–
        38, 40–41, 273, 390–392, 419–422
    after visual deprivation, 390–392
Formboard, *see* Test tray
Freezing response to aversive stimuli,
        412, 443
Frontal-lobe damage, *see also* Brain
        damage
    in aging, 461
    alternation, 153
    delayed response, 133–134, 137–138,
        143, 145–149, 151–154, 461
    and delayed reward in discrimination,
        149
    go—no-go alternation, 184
    go—no-go delayed response, 185
    hyperactivity, 518
    oddity learning, 461
Frustration, 85–86, 149–150

G

Galago, olfactory exploration by, 465
Gelada, habitat, 532
Generalization, *see* Stimulus generaliza-
        tion
Gestures, *see* Visual communication
Gibbon, *see also* Siamang
    alternation, spatial, 183
    competition for food, 552
    delayed response, 133, 140
    diet, 541
    double alternation, 267
    food preferences, 10
    group stability, 552
    group structure, 550
    habitat and distribution, 533, 535
    home range, 536

Gibbon—*Continued*
    interactions with orangutan, 541
    intergroup relations, 552, 557
    object discrimination, 183
    ontogeny of dominance, 409
    play, 409, 585
    population density, 536
    sexual behavior, 571
    skeletal fractures, 529
Go—no-go alternation, 184–185
Go—no-go delayed response, 184–185
Go—no-go discrimination, *see* Successive
        discrimination
Gorilla
    communication, 556–558
    diet, 535, 541
    discrimination learning set, 435
    dominance, 558, 567
    group stability, 552
    group structure, 550
    habitat and distribution, 533, 535
    home range, 536
    intergroup relations, 553
    investigative behavior, 466
    maternal behavior, 582
    patterned-strings performance, 439
    population density, 536
    sexual behavior, 571
    tool-using, 545
Grasp reflex, 290–291, *see also* Clinging
        by infants, Manipulation
Green monkey, *see* Vervet
Grice-type discrimination apparatus,
        414–415
Grivet, *see* Vervet
Grooming, *see also* Self-grooming
    arousal level and, 350–351, 356–357
    bonnet macaque, 565
    chimpanzee, 338, 340–341, 343–344,
        350–351, 356–357
    Japanese macaque, 531
    langur, 554
    rhesus macaque, 306–307, 543, 576–
        577
    savanna baboon, 555, 574
Group
    as social unit, 547–548
    stability, 551–552
    structure, 548–551
Guenon, 532, 557, 559, *see also* Vervet

Radiation effects—*Continued*
discrimination learning, 511–519
discrimination reversal, 509, 515–516, 518
distractibility, 503, 507–509, 511–514, 518
drug effects and, compared, 501–502, 505
elevator detour performance, 506, 517–518
extinction of object discrimination, 515
food preferences, 507
learning, 509–519
locomotion, 501–502
manipulation, 500, 503–506
oddity learning, 511, 515
pain, 520–521
patterned-strings performance, 504–505, 507, 519
reduced-cue performance, 511–512
sampling of cues, 503, 507–509, 512–516
selective death of subjects, 505, 511
self-grooming, 501–502
social behavior, 501, 508, 518
stimulus perseveration, 515–516
stimulus preferences, 508
strength of pull, 503, 507
transfer of training, 509, 512–517
transposition of intermediate-size discrimination, 516
visual acuity, 514–515, 518
visual exploration, 508, 518
win-stay-lose-shift with respect to position, 516
Random-pattern discrimination learning set, 16–18
Random responding, *see* Residual hypothesis
Range, 535–537
Reaching, work demanded in, 11–12
Recognition
perception distinguished from, 370
of surrogate mother, 388–389
after visual deprivation, 367, 374–375, 388–389
Reduced-cue problem, 510–512
Reflex stage
of infant-mother affection, 289–291

Reflex stage—*Continued*
of peer affection, 311
of sexual behavior, 323
Reinforcement, *see* Excitation, Reward, Schedules of reinforcement, Secondary reinforcement
Reinforcing function of outcomes, 71–73, 125
Rejection of infant, 303–305, 308–309, 572, 580, 582
Relational discrimination, *see* Similarity and dissimilarity problems, Transposition
Repeated discrimination reversal, 77, 177, 252, 260–261, 418
Repeated extinction and reconditioning, 251–252
Reserpine, 138
Residual hypothesis, 102, 106, 119–123
Response as cue in discrimination learning set, 81–82
Response-contingent presentation of discriminanda, 30–42
Response sequences in discrimination learning set, 99–100
Response-shift error factor, 66–67, 79–81, 85, 98
Response-shift learning set, 81–82, 182–183
Response-shock interval, 213
Restle's mathematical model of discrimination learning set, 89–91
Retention
classical conditioning, 411–412
delayed response as test of, 131–135, 138, 155, 270
infant-mother affection, 298–299
Konorski test, 190–193
object discrimination, 64
ontogeny of, 443
Reunion after social separation, 307–308, 350–351, 356–357
Reversal, *see* Discrimination reversal
Reward, *see also* Outcomes, Reward frequency, Reward quality, Reward quantity
auditory, 473–474, 486–488
delayed, 3, 149, 413
intermittent, 73, 235–240
manipulative, 413, 468–470, 480–481, 486–488

SUMMARY OF PRIMATE CLASSIFICATION

Suborder	Superfamily	Family	Subfamily	Genus	Vernacular name(s)
Prosimiae (prosimians)	Tupaioidea	Tupaiidae	Tupaiinae	*Tupaia* *Dendrogale* *Urogale*	Tree shrew Smooth-tailed tree shrew Philippine tree shrew
			Ptilocercinae	*Ptilocercus*	Pen-tailed tree shrew
		Lemuridae	Lemurinae (greater lemurs)	*Lemur* *Hapalemur* *Lepilemur*	Lemur Gentle lemur Sportive lemur
	Lemuroidea		Cheirogaleinae (lesser lemurs)	*Cheirogaleus* *Microcebus*	Dwarf lemur, mouse lemur
		Indriidae	Indriinae	*Avahi* *Propithecus* *Indri*	Woolly lemur Sifaka Indri
		Daubentoniidae		*Daubentonia*	Aye-aye
	Lorisoidea	Lorisidae		*Loris* *Nycticebus* *Arctocebus* *Perodicticus*	Slender loris Slow loris Angwantibo Potto
		Galagidae		*Galago*	Galago (bush-baby)
	Tarsioidea	Tarsiidae		*Tarsius*	Tarsier
		Callithricidae		*Callithrix* *Leontocebus*	Marmoset Tamarin, pinche